ELECTRONICS FOR ELECTRICAL TRADES

FOURTH EDITION

ELECTRONICS FOR ELECTRICAL TRADES

FOURTH EDITION

JAMES F. LOWE

B.Ed.(TAFE), Dip. Teach., I.Eng., F.I.Elec.I.E., F.A.I.E.A.

Formerly Head of Division, Electrical and Instrument Trades,
School of Applied Electricity, NSW Department of Technical and
Further Education

McGraw- Hill Book Company Sydney
New York St Louis San Francisco Auckland Bogatá
Caracas Hamburg Lisbon London Madrid Mexico Milan
Montreal New Delhi Oklahoma City Paris San Juan
São Paulo Singapore Tokyo Toronto

First published 1971
Reprinted 1972, 1973, 1974, 1975, 1976

Second edition 1977
Reprinted 1978, 1979, 1980, 1981, 1982

Third edition 1983
Reprinted 1984, 1985, 1987 (twice)

Fourth edition 1989

Reprinted 1990 (twice), 1992 (twice), 1994, 1995

A52608

National Library of Australia
Cataloguing-in-Publication data:

Lowe, James F. (James Fred).
  Electronics for electrical trades.

  4th ed.
  Includes index.
  ISBN 0 07 452608 1.

  1. Electronics. I. Title.

621.381

Produced in Australia by McGraw-Hill Book Company Australia Pty Limited
  4 Barcoo Street, Roseville, NSW 2069
Typeset in Australia by Midland Typesetters Pty Ltd
Printed in Australia by McPherson's Printing Group

Sponsoring Editor: Gillian Souter
Production Editor: Paul Cliff
Designer: George Sirett
Technical Illustrators: Colin Bardill and Diane Booth

Contents

Preface to the fourth edition

Although this book retains the same name as the three previous editions, it presents the subject of electronics (pertaining to the electrical trades) in an entirely different manner. The change in presentation, emphasis and material is due to the requirement that educational institutions meet the needs and desires of industry.

This book specifically follows the syllabus laid down by the NSW Department of Technical and Further Education for students in Stage 3 of the Electrical Trades Course. It does, however, go beyond the syllabus in certain areas, so that a wider general knowledge of electronics can be gained by the reader. As an example, on the subject of amplification the characteristics and applications of the operational amplifier are covered, as well as the application of voltage regulators, timers and logic.

There is a summary at the end of each unit, but no review questions or numerical problems have been included. Problems and theory questions pertaining to each unit have been included in a companion volume entitled the *Electronics for Electrical Trades Workbook*. In addition, practical 'hands on' exercises make the workbook an extremely valuable complement to this book.

Calculation examples in this edition use only fundamental units, and answers are accurate to only three significant figures. Moreover, any resistance values in answers have been brought to the nearest E12 preferred value.

Extra reading material has been included at the end of some units; other material which may be useful has been included in appendixes at the end of the book.

Being attuned to the requirements of present and future electrical tradespeople, this book and its companion workbook will be valuable aids in the teaching of electronics to electrical trades students.

Acknowledgments

I wish to thank Gary Renshaw (Head of Division of Electrical and Instrument Trades, School of Applied Electricity, NSW Department of Technical and Further Education) and Keith Brownlea (Head Teacher of Electrical Trades, Newcastle Technical College) for their advice and assistance in the preparation of this fourth edition.

Thanks are also extended to the following organisations for technical information and the supply of equipment or photographs:

Dick Smith Electronics
Electronics Australia Magazine
ICL Australia Pty Ltd
National Semiconductor (Australia) Pty Ltd
Philips Electronic Components and Materials
Philips Scientific and Industrial
Power Electronics, A Division of Warman
    International Ltd
Siemans Ltd
State Rail Authority of NSW
Tomago Aluminium Company Pty Ltd

Waveforms and the CRO

1.1 Alternating currents in electronics

An electrical supply, or signal voltage, is said to be alternating when it periodically changes polarity in a regular manner. Although we are familiar with *alternating* currents—being the usual form of our power supply—many *forms* of alternating currents are used in electronic circuits, and for different purposes.

The power supply from the mains is invariably in a *sinusoidal* waveform. This is for a variety of reasons, some of which are that the sine form is:

- a fundamental waveform of nature;
- a simple one to treat mathematically;
- the only waveform that can be passed through a transformer without changing its shape.

However, many other types of waveforms are used in electronic circuits—of signal, control and power types.

Figure 1.1 shows, beside the sine-wave, three other types of commonly encountered waveforms.

Later, in Units 3 and 4, we will consider rectification, where the alternating sine-wave of the supply is changed to a pulsating *direct* current and all calculations are simplified as they refer only to the sinusoidal waveform.

1.2 Alternating current characteristics

The characteristic feature of a regular alternating current is that it consists of a waveform which continually repeats itself about a *zero* reference-axis. (As it is usually considered symmetrical about the axis, the average value is *zero* for a *complete cycle*.)

A number of terms are used to describe the features of alternating waveforms. Some of these are:

Cycle One complete set of positive and negative values of a symmetrical wave.

Periodic time (T) The time taken to complete one cycle, usually measured in seconds (or electrical degrees).

Frequency The number of cycles completed in one second. Frequency is measured in *hertz* (Hz).

Average value The average value for a *complete* alternating cycle is *zero*.

Effective or RMS (root-mean-square) value The rms value of an alternating current is that value of the waveform which will produce the same heating effect as a steady state dc value.

Peak values The maximum *instantaneous* value a wave reaches in both positive and negative directions. (Note that for a square wave this value is maintained over each half-cycle.)

Peak-to-peak values The value measured from the positive peak to the negative peak. (Or, for a symmetrical wave, twice either peak value.)

1.3 Unidirectional (dc) waveforms

As the definition of an alternating waveform is one that changes polarity periodically in a regular manner, it follows that a unidirectional waveform is one that at *no time* changes polarity. (It may vary in amplitude, that is height, with respect to time, in a regular or irregular manner—but current is *always* in the one direction.)

Any variation in amplitude of a unidirectional waveform is termed *ripple*. The *average* value of any unidirectional waveform is the steady-state current

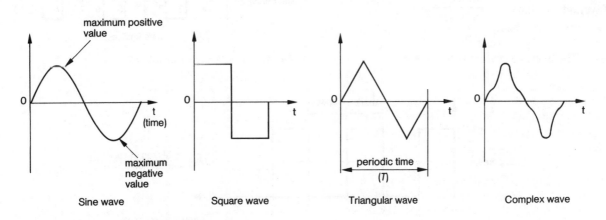

Sine wave　　Square wave　　Triangular wave　　Complex wave

Fig. 1.1　*One complete cycle of four different waveforms of alternating current*

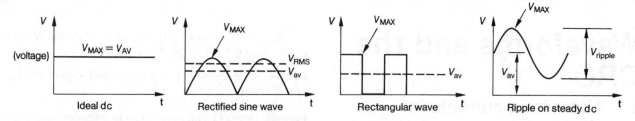

Fig. 1.2 *An ideal or steady state dc, and three unidirectional dc waveforms*

equivalent value. 'Steady-state', or 'ideal' dc has no ripple and does not vary with time. Figure 1.2 illustrates three different unidirectional waveforms with a steady-state dc.

Average value The average height or level represents the dc component of the wave and is dependent on the shape of the wave.

Ripple The time-varying component of the wave and its magnitude is the peak-to-peak value of the component.

Ripple frequency The number of ripple pulses per second.

RMS value The value of any waveform which produces the same heat as a steady-state dc.

The average value of a unidirectional waveform is that value which is registered on a moving coil instrument. For a rectangular wave it is a simple matter to determine the average height without resorting to instruments.

Example 1.1

Determine the expected reading on a moving coil instrument when measuring the current of the waveform in Figure 1.3.

Over three complete cycles (a periodic time of 6 ms)

$$\text{total area} \ = \ 10 \text{ As} \times 3 \times 10^{-3}$$

$$= \ 30 \times 10^{-3} \text{ As}$$

Now as total periodic time is 6 ms

$$I_{AV} \text{ (average current)} \ = \ \frac{30 \times 10^{-3}}{6 \times 10^{-3}}$$

$$= \ 5\text{A}$$

Answer: The average current is 5 amperes.

1.4 Periodic current reversal (PCR)

It is possible to have an *asymmetrical ac* waveform. This is a waveform in which the positive peak value and negative peak value are different. Over the period of the waveform there is a reversal of polarity but, depending on the waveform, there may only be a small excursion into the alternate polarity. Figure 1.5 shows a sinusoidal PCR waveform. The average value of this waveform is not zero (unlike a symmetrical, or 'normal' ac wave) but has a value dependent on the waveform.

As there *is* a current reversal each regular period it is referred to as a PCR waveform. Consider the waveform in Figure 1.4. This is an asymmetrical, or PCR, mark-space rectangular waveform. As it is rectangular it is easy to determine its average value without using complicated mathematics.

Example 1.2

Determine the average value (the value read on a moving coil instrument) of the current in the waveform represented in Figure 1.4.

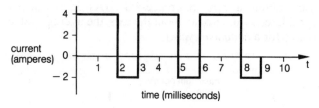

Fig. 1.4 *Waveform for Example 1.2*

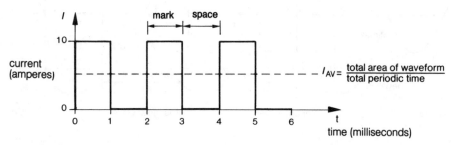

Fig. 1.3 *Waveform for Example 1.1*

Over three complete cycles (a periodic time of 9 ms),

$$\text{total area (positive)} = 4 \times 2 \text{ As} \times 3$$
$$= 24 \text{ As}$$
$$\text{total area (negative)} = -2 \times 1 \text{ As} \times 3$$
$$= -6 \text{ As}$$
$$\text{total area (net)} = 24 - 6$$
$$= 18 \text{ As}$$

Now as total periodic time is 9 ms,

$$I_{AV} = \frac{18}{9}$$
$$= 2A$$

Answer: The average current is 2 amperes.

1.5 RMS value of unidirectional waveforms

Earlier it was quoted that the rms value of a wave is that value of the wave which will produce the same heat as a steady direct current. Mathematically it is the square root of the squares of the means (averages) of that particular waveform. The usual instruments used to measure dc values are moving coil instruments, or digital instruments which simulate moving coils, and so these readings always present average values.

For magnetic circuits or electrolysis it is the average values that must be considered, but for heating effect it is the rms value which is important. In a dc electric motor the average value is performing the work of the motor but it is the rms value which produces heating in the motor windings.

1.6 Summary of waveforms

In electronic circuits electrical energy or signal voltages can appear as many diverse regular waveforms. If these are symmetrical about an axis they are referred to as alternating voltages. If they are of a varying nature, either random or regular, and at no time does the polarity reverse, then they are referred to as unidirectional voltages.

Voltages which are asymmetrical about a zero axis, usually in a regular manner, are referred to as PCR voltages.

The average value of an alternating voltage is zero, but the average value of a unidirectional or PCR voltage is the area under the curve (considering both positive and negative) divided by the time—either for a cycle or a total number of cycles. This is the value that is read on a moving coil, or equivalent, instrument.

The rms value of an alternating, PCR or unidirectional voltage is that value which produces heat equivalent to the same value of a steady voltage.

1.7 Observing waveforms

Whilst we may be able to measure average and also rms values of alternating, PCR and unidirectional waveforms (and even, with some instruments, the peak-to-peak values), unless it is known that we are dealing with a regular waveform, such as the mains supply, we have no idea whatsoever of the *form* of the wave.

The only instrument which may be used to observe a waveform is the oscilloscope. Early oscilloscopes were mechanical and used revolving mirrors and mirrors attached to coils to produce an observable waveform. These mechanical monsters are now relegated to museums and almost all modern oscilloscopes use cathode ray tubes as the viewing devices. For this reason they are referred to as *cathode ray oscilloscopes*, invariably shortened to CRO.

1.8 Cathode ray tubes

A cathode ray tube belongs to a special class of *vacuum tubes* which use beams of electrons, in a vacuum, produced by a *heated cathode*. The electrons are produced by *thermionic emission*.

The cathode ray tube consists primarily of a cathode and a *phosphor screen*, together with electrodes which are able to focus and deflect the electrons emitted from the heated cathode. The simplified internal construction of a cathode ray tube is shown in Figure 1.6.

Referring to Figure 1.6 we find on the left side a *heater* which, when connected to a low voltage source

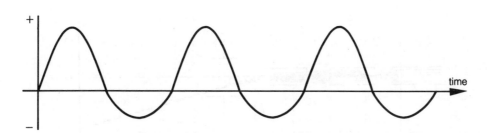

Fig. 1.5 *A sinusoidal PCR wave. Note that although the positive and negative amplitudes are different, they both follow half of a sine curve*

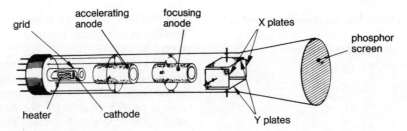

Fig. 1.6   *Internal construction of a cathode ray tube*

(usually 6.3 volts), glows at a dull red heat. Surrounding the heater, and insulated from it, is a small tube called the cathode. The cathode tube is coated with rare earth oxides which will provide a copious emission of electrons when heated. Another cylinder with one end partially opened surrounds the cathode. This is termed the *grid* and when a suitable negative potential is applied to this it restricts the emission to a thin beam.

The phosphor screen of the tube is made conductive and is supplied with a high, positive potential (in the larger tubes it can be many thousands of volts), which attracts the electrons from the cathode. Because the interior of the tube is sealed and operates at a very high vacuum there is no impediment to the flow of electrons across the space.

To assist the flow of electrons from the cathode to the phosphor screen, another larger cylinder called the *accelerating anode* (open at both ends) is placed in front of the grid. This has a positive potential applied to it, but less than that of the phosphor screen, and attracts the negative electrons. A further cylinder called the *focusing anode*, similar to the first and again in front of it, also has a positive potential. This further attracts the electron stream, in such a way that the stream is converged to a small spot when it arrives at the phosphor screen at the face of the tube. (Both these anodes behave in a similar manner to a lens and a beam of light.)

The only *control electrodes* of a simple cathode ray tube are the *deflection plates* for horizontal deflection (named *X plates*) and for vertical deflection (*Y plates*). By applying suitable potentials to these plates, the beam of electrons may be focused at any position, as a small spot on the face of the tube. The phosphor on the face of the tube glows when under the influence of the beam. This type of deflection is termed *electrostatic deflection*. Figure 1.7 illustrates the beam formation, focusing and

deflection, from cathode to phosphor screen in the cathode ray tube.

1.9 Uses for cathode ray tubes

Although challenged by *liquid crystal displays* in some applications, notably portable lap-top computers, the cathode ray tube is still supreme for a highly visible crisp and clear display. The cathode ray tube forms the screens of CROs (see Fig. 1.8), the picture tubes of TV receivers, the playing fields of video games and the output displays of computers and data transmission terminals.

No matter what application a cathode ray tube has, its principle of operation is the same. The actual physical construction of the tube, however, differs depending on the use. Monochrome (black and white) TV picture tubes have a large area of phosphor and use magnetic deflection for the electron beam. Monochrome (green or amber) computer *visual display units* use a very high-quality phosphor and special deflection circuits. Colour TV picture tubes use three electron guns and a phosphor made up from a multitude of tiny segments of the three primary colours, each excited by only one of the electron beams; the highest quality computer colour VDUs (visual display units) use a similar but much higher quality construction.

1.10 The cathode ray oscilloscope

The CRO is used to observe waveforms of time-varying signals and related data. CROs differ slightly in appearance in the simpler types but their controls all perform much the same function. There are some very

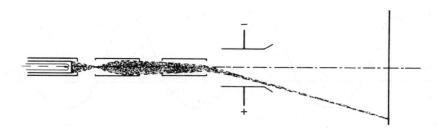

Fig. 1.7   *Focusing and deflection of electron beam*

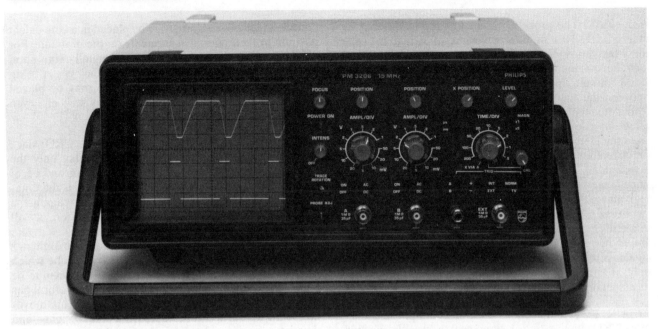

Fig. 1.8 *A cathode ray tube forms the viewing screen of a CRO* PHILIPS SCIENTIFIC AND INDUSTRIAL DIVISION

sophisticated CROs but they are not within the scope of this book.

The simpler CROs all have these features:

- a viewing screen fitted with a graticule;
- a shielded and an 'earth' input terminal;
- an ac or dc input coupling;
- a vertical or Y amplifier, with stepped attenuation;
- a vertical shift;
- a time-base with fixed and variable control;
- a horizontal or X shift;
- a horizontal gain control;
- a trigger level and trigger sensitivity control
- a brightness control;
- a focus control.

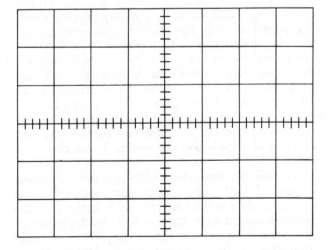

Fig. 1.9 *A CRO screen graticule*

1.11 Controls of the CRO

As a CRO is a reasonably complex device it is wise to refer to an instruction manual or receive guidance before attempting to operate it. However, once a person is proficient in operating a particular CRO it becomes relatively simple to operate another—like a driver adapting to different motor vehicles. The controls are shown in Figure 1.10.

In this section the general features of a CRO will be outlined but it must be remembered that the layout of controls will differ with particular makes and many CROs have special features which may or may not be required. Reference will be made to the features listed in Section 1.10.

Viewing screen The viewing screen, as previously explained, is the phosphor screen of the cathode ray tube. On this screen the beam of electrons traces the waveform to be observed. The phosphor of the screen glows (usually green) when struck by the electron beam. In front of the screen is placed a *graticule*, a piece of plastic material (usually transparent green to improve the contrast) engraved in centimetre divisions. (These divisions are of assistance when viewing a waveform and are essential when measuring it, as will be discussed later.) A full-size representation of a CRO screen graticule is shown in Figure 1.9. The fine engraved lines of the graticule are often filled with a red dye which, when side-lit by a lamp with an adjustable intensity, highlights or diminishes the effect of the graticule lines.

Input terminals The input terminals consist of two types, a *shielded input* and an 'earth' input. The shielded input has a socket into which a *probe lead* is connected. (Since this lead may be connected to high impedance circuits, which could be affected by stray induced voltages, it is shielded.) The probe at the end of the lead could be marked 1:1, 10:1 or have a switch, to switch between these two ratios. All this means is that the incoming signal may be *attenuated* or reduced by the particular ratio. If the circuit under test was of very low impedance between the two connected points, a simple unshielded lead could be used; however, as the shielding has no effect on a low-impedance circuit, it is usual to always use a shielded lead. The 'earth' lead connects to a common or reference point in the particular circuit being investigated. The *input impedance* of the CRO (the impedance between the two input leads) is usually quite high and may vary from 2 megohms to 10 megohms.

Some CROs may have two shielded input terminals marked channel 1 and channel 2 (or A and B). These are known as 'two-channel' CROs and may display two signals simultaneously. If observing a single waveform only, it is necessary to use only one of the inputs and its associated controls.

Ac or dc input Adjacent to the shielded input terminal, most CROs have a two-position switch (usually marked 'AC' and 'DC'). In the DC position the input connects directly to the internal circuits, but in the AC position a capacitor is connected between the input and internal circuitry. When switched to DC, the CRO will respond to any dc level that may be present in a signal—in other words, it will display the signal with reference to zero potential. When switched to AC, any dc component in the signal is ignored and the signal waveform is displayed about the set mean position on the screen. When observing an ac waveform, the display will not be affected by the switch position. However, when observing a uni-directional waveform with a ripple (similar to that on the right side of Figure 1.2), switching the input to AC will only display the ripple and will ignore the dc content below it. If the ripple content is high it is important that this facility is available, but it must be stressed that this effect is similar to zero suppression in a graph.

Vertical (or Y) amplifier The vertical amplifier usually has two controls, which may be concentric. They are the volts/cm (or *ampl/div*) switch and the *position* (or *vertical shift*) control. These controls, also, are usually placed adjacent to the shielded input terminal. The volts/cm switch may have twelve positions, each marked with a number representing the voltage of the waveform per centimetre (or between divisions on the graticule). The range of these voltages could be 0.01–50 volts. This switch can magnify or diminish the vertical size of the displayed waveform. (It must be remembered that if using a 10:1 probe, all these readings must be multiplied by 10; some CROs may have a 1:1 and 10:1 switch rather than using an attenuator probe.) The internal output of the vertical amplifier is connected to the Y plates of the cathode ray tube and causes the beam to be deflected vertically in accordance with the instantaneous values of waveform to be observed.

The 'position' (or vertical shift) control can shift the observed waveform vertically on the screen. This is often done so that the waveform may be placed in a convenient position to facilitate measurements on the graticule. For example an ac wave could be centred equally above and below the horizontal centre graticule line. When using two inputs on a two-channel CRO one may be placed above another for comparison purposes or even superimposed for the same reason.

Time-base The time-base is the heart of a CRO, since without this feature the screen could display only the maximum peak value of an input waveform and be of no more value than an analogue voltmeter. It is the time-base that is able to display the waveform against a horizontal (or *x*-axis) with a time calibration, as all waveforms are a function of time.

The time-base circuit consists of a variable oscillator connected to the X plates of the cathode ray tube which produces a saw-tooth waveform causing the beam to be deflected from side to side. The saw-tooth waveform has a uniform rise and then a very quick fall. This moves the beam across the tube face, from left to right—and then at a very much relatively faster rate returns the beam to the left again. Special circuits cut off the beam when the return is taking place. This backward and forward movement of the beam will produce a single horizontal line across the screen as the phosphor is excited. To show the waveform on the screen, the Y plates must deflect the beam up and down. This signal must vary exactly as the voltage of the measured source, and since it is moving the beam across the screen at the same relative speed it traces exactly the waveform of the source.

The screen phosphor is made so that there is a slight persistence of emitted light on the screen after the beam has moved on. This, with the normal persistence of the eye's vision, enables a steady trace to appear on the screen. (It should be remembered that this trace is made up by a very rapidly moving spot, as indeed is any display on a cathode ray tube screen.)

The time-base control is usually in two parts, a *time/cm* (or time/div) switch, and a *variable* (or *vernier*) control, which are sometimes mounted concentrically. The time/cm switch could be a twenty position switch with calibrations from 500 milliseconds/centimetre to 1 second/centimetre and an OFF position. The time-base switch position is chosen so that the observed waveform will display, across the screen, just enough cycles for correct observation. This is at the discretion of the operator. Whatever time-base position is chosen, it is possible to determine periodic time over a cycle by observing the distance in centimetres between the start and finish of a cycle and multiplying by the time-base setting.

The variable control has a position marked 'calibrate' (or possibly 'CAL') and only in this position does the time/cm reading of the time/cm switch hold true. If the control is moved from this position, the frequency of the time-base may be altered to obtain a better observation of the waveform, but no longer can the periodic time be determined from the screen.

Fig. 1.10 *Controls of a CRO*   PHILIPS SCIENTIFIC AND INDUSTRIAL DIVISION

Horizontal control The horizontal control (sometimes termed the X amplifier) can again consist of two concentric controls, which can be termed *horizontal position* (or *horizontal shift*) and *horizontal gain* (or *horizontal magnification*, often abbreviated to HORIZ MAG). The horizontal shift control can move the observed waveform back and forth across the screen, so that the commencement of a cycle can be positioned at a graticule mark to make measurement on the screen face easier when determining the periodic time of a wave.

The horizontal gain or magnification control has a calibration position (again for accurate determination of periodic time), and when moved from this position can expand the waveform across the screen (often up to five times enlargement) to more closely study a particular part of the observed waveform. This control may appear similar to the variable control on the time-base, but its operation depends on different internal circuits, and its operation can be made completely independent of the time-base under certain circumstances.

Trigger control The trigger control is necessary to synchronise the time-base to the incoming waveform. Without the trigger circuits (or if they are incorrectly set), the observed waveform could appear as an unrecognisable jumble on the screen. The trigger control

may consist of two parts: *trigger level* and *trigger sensitivity*. The trigger level depends on what part of the waveform the observed wave will commence being traced. As well, it usually has a position marked 'auto' when no particular level is required. The trigger sensitivity adjusts the trigger control for differing levels of incoming signal, to ensure reliable triggering.

Brightness and focus The *brightness* control is similar to that on a television receiver and can be made to adjust the brightness of the waveform trace on the screen. Some observed waveforms may require adjustment of the brightness control to get an easily-seen trace on the screen.

The *focus* control is used to obtain a clear sharp trace on the screen. On some CROs a change in brightness requires an alteration of the focus control.

1.12 Using the CRO

We refer now to a few of the most common uses of a CRO. The first is the measurement of a dc voltage.

If the time-base is set to some convenient setting (such as 5 ms/cm) and the leads are joined together, a single line trace will appear across the screen. This can be moved vertically by the Y shift control and positioned on a horizontal line of the graticule. Now, if the leads are placed across a dc potential (and the V/cm switch is on a suitable setting), the line will move vertically up or down depending on the polarity of the leads position: if the shielded lead is placed on the positive side of the potential the trace line will rise, if the leads are reversed it will fall.

Example 1.3

Figure 1.11 represents the screen of a CRO. The dashed line represents the position of the trace when the two leads are shorted together. The solid line represents the trace when the leads are placed across a dc potential. If the volts/cm switch is set to the 0.5 position, what is the potential across the CRO leads?

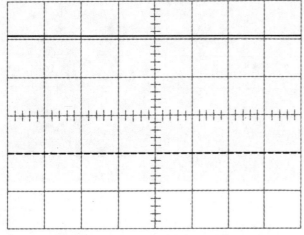

Fig. 1.11   *CRO screen representation for Example 1.3, a dc voltage*

The vertical distance between the shorted position and the measuring position is 3.1 cm. As the volts/cm switch is set to 0.5 volts/cm

measured voltage   = 3.1×0.5

                   = 1.55 V

Answer: The potential measured by the CRO is 1.55 volts.

The voltage measurement above could just as easily have been measured with a voltmeter—but with ac waveforms not only can we observe the form of a wave, we can also measure its peak value and determine its frequency as well.

Example 1.4

Consider the waveform represented in Figure 1.12. If the volts/cm switch is on the 2 position, a 10:1 attenuator probe is being used, the time-base is set to 1 ms/cm, and all other relevant controls are on the CAL position—what are the peak-to-peak voltage, the rms voltage and the frequency of the observed waveform?

Since the waveform has been vertically positioned equally above and below the centre-line of the graticule and also so that a zero point on the wave is in line with the first vertical centimetre line of the graticule, measurement from the face of the screen is made easier.

From the screen, it can be seen that the peak-to-peak measurement is 2.6 cm and the width of one cycle is 2.65 cm. From these measurements the answers can be calculated.

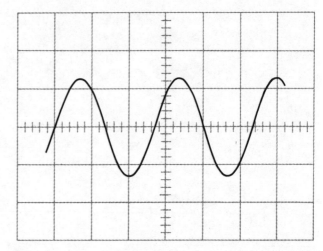

Fig. 1.12   *CRO screen representation for Example 1.4, a sine wave*

Now as the volt/cm setting is 2 V/cm and the attentuation of the probe is 10:1, each centimetre on the face of the tube represents

2×10 = 20 V

As the peak-to-peak (p-p) distance is 2.6 cm
   p-p voltage   = 20×2.6
                 = 52 V

Now as it is obviously a sine wave, the positive peak voltage is half the p-p voltage

   peak voltage   = $\frac{1}{2} \times 52$
                  = 26 V

Also because it is a sine wave, the rms value is 0.707 of the peak value, so

   rms voltage   = 0.707×26
                 = 18.4 V

To determine the frequency (f) of the waveform we first find periodic time (T), and as the time-base is set to 1 ms/cm and the measured distance of the period or cycle is 2.65 cm

   T = 0.001×2.65
        = 2.65×10^{-3} s

Now as frequency equals the inverse of periodic time ($f=1/T$);

$$f = \frac{1}{2.65 \times 10^{-3}}$$
$$= 377 \text{ Hz}$$

Answer: The peak-to-peak voltage is 52 volts; the rms voltage is 18.4 volts; and the frequency is 377 hertz.

Example 1.5

Figure 1.13 represents an observed waveform on the screen of a CRO. The volts/cm switch is set to 5, a 10:1 probe is being used, the time-base is set to 5 ms/cm, and the other X controls are on CAL. What is the amplitude of the ripple and what is the ripple frequency?

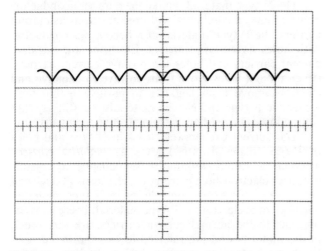

Fig. 1.13 *CRO screen representation for Example 1.5, a ripple on a dc voltage*

From the screen we can measure the amplitude of the ripple to be 0.3 cm and the period of the ripple to be 0.6 cm. Now using the settings given above

$$\text{ripple amplitude} = 5 \times 10 \times 0.3$$
$$= 15 \text{ V}$$
$$T = 5 \times 10^{-3} \times 0.6$$
$$= 0.003 \text{ s}$$
$$f = \frac{1}{0.003}$$
$$= 333 \text{ Hz}$$

Answer. The ripple has an amplitude of 15 volts and a frequency of 333 hertz.

Unit 1 SUMMARY

- Mains supply ac is a sinusoidal waveform, but in electronics many other waveforms may be encountered.
- When dealing with alternating waves, the terms cycle, periodic time, frequency, average value, rms value, peak value and peak-to-peak value must be understood.
- The amplitude of a dc voltage may vary with time, and the terms average value, ripple, ripple frequency and rms value must be understood.
- A cathode ray oscilloscope (CRO) is an instrument for the observation and measurement of waveforms.

- The viewing screen of a CRO is the face of a cathode ray tube.
- A graticule is fitted to the screen of a CRO to assist in the accurate measurement of an observed waveform.
- The input probe lead of a CRO is shielded to prevent stray induced voltages from interfering with the observed waveform.
- The CRO input can be switched from ac to dc to observe a waveform with a reference to zero.
- The vertical or Y amplifier of a CRO determines the height of the observed waveform on the screen.

Diodes

2.1 The perfect diode

The word 'diode' comes from two Greek words: *dia* (through) and *hodos* (a way). In its simplest sense a diode is an electronic device which will only allow current to pass through it in one direction. We say current 'flows' from the positive terminal (or *anode*) to the negative terminal (or *cathode*) within the device.

In a diode, the terms anode and cathode are applied to distinct terminals of the device, such that unless the anode terminal is positive and the cathode terminal is negative, there will be no current through the device. In other words, it must have the correct polarity applied before it will conduct. In this sense the diode may be thought of as a 'switch' which is 'on' when the anode has a positive potential applied and 'off' when the anode has a negative potential applied, with respect to the cathode.

Figure 2.1 shows the voltage-current characteristics of a perfect diode. It can be seen from the graph that when the applied anode voltage is negative no current flows, and when the anode is positive, current flows with no voltage drop across the device. It must be emphasised here that this is the characteristic of a 'perfect' diode and in fact no perfect diode exists, though as we will see later, in many practical cases the diode may be considered perfect.

2.2 P-type and N-type material

P-type material may be made by adding a small proportion of a three-valance element, such as indium, to a pure semiconductor, such as silicon. N-type material

may be made by adding a small proportion of a five-valence element, such as antimony, to a pure semiconductor material, such as silicon. The pure semiconductor material is then said to be *doped*. (Another semiconductor sometimes used is germanium.)

The N-type material (unlike the pure semiconductor) now becomes a conductor with free electrons as current carriers; the P-type material also becomes a conductor, with holes as current carriers. Electrons are negative current carriers and holes are positive current carriers. (Even though holes move from positive to negative and electrons move from negative to positive within their respective P-type and N-type materials, we still say that *conventional current flow* is from positive to negative.)

The current carriers (electrons in N-type and holes in P-type material) are known as *majority carriers*. Thermal agitation within the semiconductor crystals produces electron-hole pairing, which means that as well as electrons there are some holes in N-type material, and also some electrons in P-type material. These carriers, opposite to the normal current carriers, are *very* much in the minority and are referred to as *minority carriers*. (It is these minority carriers, which produce leakage current, that contribute towards the practical diode not being perfect.)

For a more detailed explanation of semiconductor theory students may refer to Appendix A.

2.3 The PN junction

When P- and N-type materials are effectively joined or *diffused*, the line of joining or diffusing creates a *junction* between the differently doped materials. This is known as the PN junction and is illustrated in Figure 2.2. It may be expected that electrons will flow from the N-type material into the P-type material to neutralise the holes, but this is not the case.

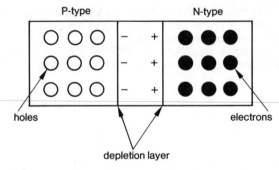

Fig. 2.2 *The PN junction. Note the negative charge on the P-type side of the junction and the positive charge on the N-type side*

There *is* a movement of holes towards the N-type region and of electrons towards the P-type region: however, as soon as the holes leave the P-type region they leave behind a negative charge, and the electrons

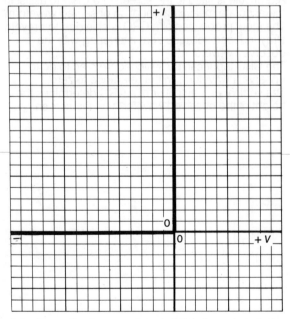

Fig. 2.1 *Forward and reverse characteristics of a perfect diode*

leave behind a positive charge in the N-type region. The produced charge then prevents any further movement of the holes and electrons. Thus there is a very narrow region at the junction, where movement of both holes and electrons have left a depleted area. Note also that across this narrow region a potential is developed and the potential is a barrier to any further movement of holes and electrons.

Various names have been given to the depleted region, including *potential barrier, transition region*, and *depletion layer*. All describe the same thing—the charge developed across the junction due to the initial interchange of holes and electrons. In this text we use the term *depletion layer*.

The charge across the depletion layer exists only at the junction; it cannot be measured by a voltmeter between the P- and N-type materials. In the theory of diode action it may be represented by an imaginary battery connected within the two crystals at the junction. See Figure 2.3.

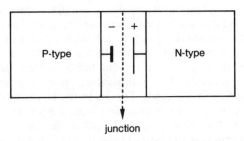

junction

Fig. 2.3 *The depletion layer can be represented as an imaginary battery*

2.4 Junction biasing

Suppose an external source of electricity is connected across P- and N-type materials with a polarity that is opposite that of the inbuilt imaginary battery or the potential across the junction (see Fig. 2.4). A voltage divider, connected so that the source can be varied upwards fom zero, and a means of measuring the intensity of current flow, should show that very little current flows when the divider is first moved up from the negative end; after reaching a certain small positive potential, however, the current rises fairly steadily as the voltage is increased (refer to Fig. 2.6).

The initial slow rise in current as the potential across the diode is increased is caused by the external supply having to overcome the potential across the junction; from then onwards the effect of the depletion layer virtually disappears. Electrons in the P-type material, near the positive terminal of the supply, break their electron pair bonds and enter the supply, thus producing new holes. Simultaneously, electrons from the negative terminal of the supply enter the N-type material and migrate towards the junction. Free electrons from the N-type then flow across the junction and move into the holes which have migrated from the positive terminal. This current, called *forward current*, will continue as long as the external supply is connected. Note that the current is produced by majority carriers and the device is said to be *forward biased*.

When the polarity of the external supply is reversed, the potential within the device is effectively reinforced and the depletion layer becomes wider. This is because the free electrons in the N-type are attracted towards the positive terminal, away from the junction, while the electrons from the negative terminal of the supply enter the P-type and migrate towards the junction. This current, called *reverse current*, is *extremely* small (see Fig. 2.5). Note that this current is produced by minority carriers and the device is said to be *reverse biased*.

2.5 The diode effect

If the reverse (or leakage) current is ignored, the PN junction device can be said to be unidirectional, or a one way current carrying device—that is, if the polarity could be reversed, only one definite polarity would allow

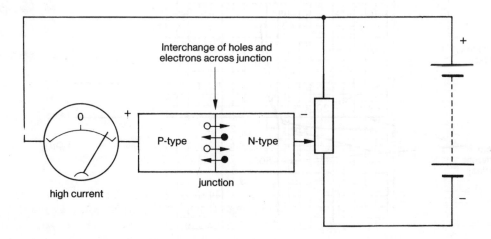

Fig. 2.4 *A forward biased junction. A high current flows*

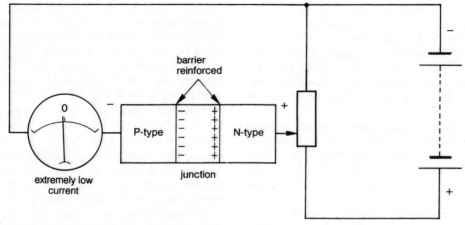

Fig. 2.5   *A reverse biased PN junction. An extremely low current flows*

conduction. This is known as the diode effect and the device is called a *PN junction diode*. (When leakage current is neglected—and it usually is, in practice—bear in mind that an *extremely* small reverse current will always exist, caused by minority carriers, and considerably increased by temperature rise.)

2.6 Practical diode characteristics

The forward and reverse characteristics of a typical junction diode are shown in Figure 2.6. Note particularly that the scale of the current axis in the negative direction is in *microamperes.*

A PN diode has a very low resistance when forward biased, and a very high resistance when reverse biased. If, however, the reverse bias voltage is increased above a certain maximum value (known as the *maximum peak inverse voltage* or PIV_{MAX}) for a particular diode, the diode may break down. In normal diodes this is exhibited as 'punch through', which destroys the diode and leaves it short-circuited and useless.

If the diode is heated above its normal maximum allowed temperature, breakdown may occur at lower than normal maximum values and again the diode may be destroyed. Heating of the diode usually occurs through excessive current or insufficient cooling at normal current. The dashed curve in Figure 2.6 shows that an increase in temperature gives a greater increase in reverse leakage current than in forward current, but both these increases may aggravate the temperature rise and lead to destruction of the diode. Diodes in the larger sizes must always be used in conjunction with a *heat sink*, which radiates the heat produced within the diode to the surrounding air and keeps the junction temperature below the maximum permissible value.

The conventional symbol for the semiconductor diode is shown in Figure 2.7. The arrow shape of the symbol points in the direction of conventional current, that is from positive to negative.

Semiconductor diodes vary in size and type (see Fig. 2.9). Two typical diodes are shown in outline in Figure 2.8, and Table 2.1 gives their normal ratings. Note the

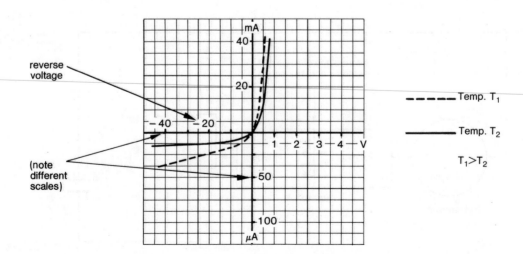

Fig. 2.6   *The forward and reverse characteristics of a PN diode. Note that forward and reverse voltage and current scales are dissimilar*

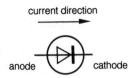

Fig. 2.7 *Symbol for a PN diode*

methods by which their respective electrodes are identified. The larger diode has the conventional symbol printed on the body. (If the symbol was reversed, it would indicate that the lead was cathode and the body was the anode.) The smaller diode uses a domed end to identify the cathode lead. An alternative method involves placing either a ring around the cathode or a dot, usually red, adjacent to it.

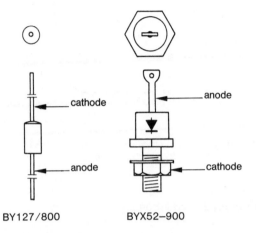

Fig. 2.8 *Two typical PN diodes, full size*

Table 2.1 *Ratings of two typical PN diodes*

Type	BYX52-900	BY127/800
Average current rating (amperes)	40	1
Maximum forward surge current (amperes)	650	10
Maximum peak inverse voltage (volts)	600	800
Maximum operating temp. (^{0}C)	125	150

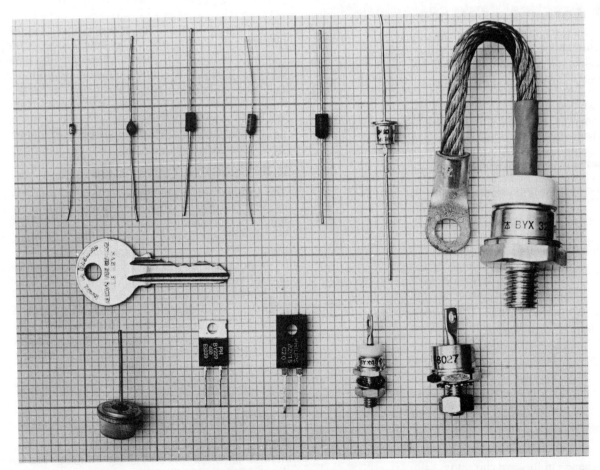

Fig. 2.9 *A range of PN diodes. Ratings vary from 200 milliamperes for the diodes on the upper left, to 150 amperes for the one on the upper right* PHILIPS ELECTRONIC COMPONENTS AND MATERIALS

Power diodes are exclusively silicon. The larger diodes carry currents in excess of 1500 amperes and others are manufactured to work with a peak inverse voltage in excess of 50 kilovolts.

Germanium diodes have been used for radio and TV but are now virtually obsolete. Their lower forward voltage drop can be seen in Figure 2.10.

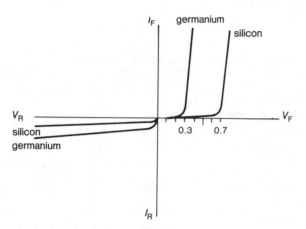

Fig. 2.10   *Combined forward and reverse characteristic curves for silicon and and germanium diodes*

2.7 Forward resistance

An examination of Figure 2.11 shows that from a forward voltage of about 0.7 volt the curve is linear, i.e. the resistance is constant. As diodes are usually designed to carry current well into the linear part of the curve, it is relatively simple to determine the forward resistance of the diode from this portion of the curve.

The curve becomes linear at point *a* and has been taken to where the voltage drop is 0.8 volt at point *b*. The change in voltage drop over the change in current determines the resistance.

This may be expressed as:

$$R_F = \frac{V_F2 - V_F1}{I_F2 - I_F1} \tag{2.1}$$

where R_F is forward resistance,
V_F2 is higher voltage drop,
V_F1 is lower voltage drop,
I_F2 is higher current,
I_F1 is lower current.

This may be expressed as:

$$R_F = \frac{\Delta V_F}{\Delta I_F} \tag{2.2}$$

where ΔV_F is change in forward voltage drop,
ΔI_F is change in current.

In Figure 2.11 the change in forward voltage drop is:
$$\Delta V_F = 0.8 - 0.7 = 0.1$$

The change in forward current is:
$$\Delta I_F = 0.5 - 0.1 = 0.4 \text{ A}$$

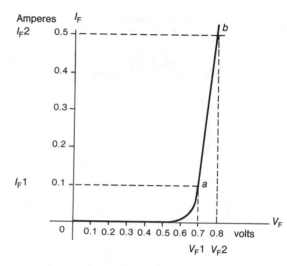

Fig. 2.11   *Determining the forward resistance of a silicon diode from the forward characteristic curve*

Substituting in equation 2.2:

$$R_F = \frac{0.1}{0.4}$$
$$= 0.25 \ \Omega$$

It should be noted that for *power diodes* the forward resistance is usually much less than 1 ohm. Compare this with the forward resistance of a common 10 ampere domestic power switch which could be 0.05 ohm, giving a voltage drop of 0.5 volt at full load. For *small signal diodes* used in communications, the forward resistance can greatly exceed 1 ohm.

2.8 Reverse resistance

The reverse resistance of a PN diode is a function of its leakage current (leakage current, moreover, is a function of the diode's temperature, see Section 2.9). As leakage current is produced by minority carriers, it is usually only very small. A typical value in a power diode could be about 20 microamperes at a reverse voltage of 100 volts. This would indicate a reverse resistance of about 5 megohms. Compare this with the insulation resistance of an opened domestic power switch, which could be 6 megohms. This figure and the forward resistance figure from Section 2.7 would indicate that a forward and reverse biased PN diode is, in effect, very similar to an open and closed switch.

2.9 Temperature effect on diodes

With an increase of temperature there is an increase of thermal agitation of the atoms forming the crystal structure of the PN diode. This releases more electron-hole pairs and produces an increase of minority carriers.

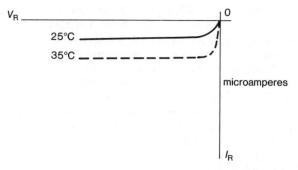

Fig. 2.12 *Increase of reverse leakage current with an increase in temperature*

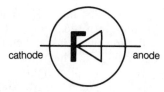

Fig. 2.14 *A zener diode symbol*

Thus leakage current in a reverse biased diode is increased. This is shown in Figure 2.12. The increase in leakage current is approximately twice for each 10°C increase in temperature. As an example, if the leakage current at 20°C was 30 microamperes, it would be 60 microamperes at 30°C.

The effect on the forward characteristics is that the forward voltage drop decreases. This is due to electron-hole pairing providing additional majority current carriers. This is illustrated in Figure 2.13. The decrease in forward voltage drop, which is actually a decrease in the depletion layer barrier potential, is about 2.5 millivolts for each 1°C. This means that for a temperature change of from 25°C to 45°C—a rise of 20°C—the forward voltage drop could fall from 0.7 volt to about 0.65 volt.

The symbol for the zener diode is illustrated in Figure 2.14. The physical appearance of a zener diode is little different from that of a normal diode and it is usually only by the type number (or perhaps the symbol printed on the body in the larger sizes) that these units can be identified.

From the characteristic curve, in Figure 2.15, it can be seen that the zener diode, when reverse biased, behaves like a normal diode until the *zener point* is reached. At this point, reverse current commences and the high reverse resistance virtually disappears (for most practical purposes), the voltage drop across the diode remaining practically constant regardless of the current (within limits).

The zener diode has this remarkable property of maintaining an almost constant voltage across it, provided the power dissipation within it is limited by external resistance. Zener diodes are invariably used in this way, so the normal connection for the device is in the reverse biased condition. These diodes are used for voltage references and as voltage regulators, and their application is investigated in Unit 7 of this book.

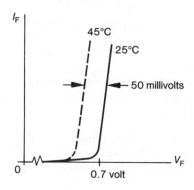

Fig. 2.13 *Decrease of forward voltage drop with an increase in temperature*

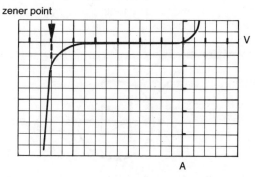

Fig. 2.15 *Typical zener diode characteristics*

2.10 Zener diodes

We have seen, in Section 2.6, that in normal PN diodes voltage in excess of the maximum peak inverse voltage (PIV_{MAX}) will destroy the junction. With careful control of doping, however, it is possible to manufacture PN diodes in which a reverse current or avalanche occurs before the punch-through condition is reached. This effect was noticed by Carl Zener in 1934 and the device based on his observations has been named the *zener diode.*

2.11 Light emitting diodes (LEDs)

In any PN diode (as in any conductor) the passage of current carriers is accompanied by a release of energy in the form of heat. In large power diodes, this heat loss may be so great that special efforts are necessary to remove the heat, so the temperature at the junction of the diode is kept below a maximum permitted level. Now if the diode is specially constructed, and uses a semiconductor compound such as gallium arsenide (GAs), the energy loss is in the form of light, as well as heat. These special PN diodes are called *light emitting diodes* or more commonly *LEDs* (see Figure 2.16).

By careful control of the material doping, the light emitted may be yellow, green, red or infrared. Some LEDs are even made with three junctions, and emit red, yellow or green light depending on which one of its leads is selected. LEDs are used as pilot lights, TV and radio channel indicators, bar graph arrays and as digital readouts. LEDs will be more fully covered in Unit 27 of this book.

2.12 Photodiodes

As discussed in section 2.3, a depletion layer is formed at a PN junction. When reverse biased, a very small leakage current flows, due to electron-hole pairing, and consequently reverse leakage current will increase rapidly with increase in temperature.

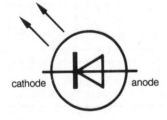

Fig. 2.16 *Light-emitting diodes (LEDs) and symbol*

Light energy falling on the junction produces the same effect as heat does. The holes and electrons freed by the action of the light energy (electron-hole pairing) are acted on by the reverse bias potential—which sweeps the holes one way and electrons the other way, and constitutes a current in the reverse bias direction. The intensity of the current is in proportion to the light intensity at the PN junction. The symbol for a photodiode, and a section view of a typical photodiode, can be seen in Figure 2.17. Applications of photodiodes will be covered in Unit 27.

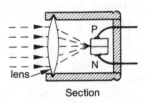

Fig. 2.17 *Symbol and section of a typical photodiode*

2.13 Diodes as current guides (steering diodes)

In section 2.1 it was explained that diodes will only allow conduction in one direction. This enables a diode to function as a *current guide* so that current may only flow in one direction in a circuit. Diodes employed in these circuits may be called *current guide* diodes or *steering* diodes. In this section we will examine four applications of such diodes.

Selective lamp switching Figure 2.18 shows two lamps connected to a double-pole two-way switch by only two wires—either lamp may be operated independently. With

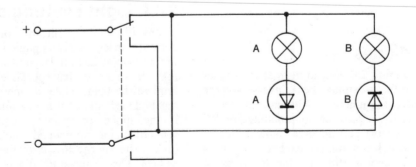

Fig. 2.18 *Current guide diodes used in selective switching*

the switch blades in the upper position, the upper wire in the diagram is made positive, and the lower wire negative. Diode A is forward biased and lamp A will light. Diode B is reverse biased and so lamp B will not light. With the switch operated so that the blades are in the lower position, the top wire now becomes negative and the lower wire positive. This reverse biases diode A and so lamp A will not light. Diode B is forward biased and so lamp B will light. Operation of the switch will alternatively light lamp A or B.

Common component operation (car warning-light application) Most cars are equipped with warning lights for such malfunctions as low oil pressure, high water temperature, and no charging current. At night these indicating lights are easily seen but during the day they can be missed. The provision of an aural warning by means of a buzzer, in addition to the lights, would diminish the possibility of the warning lights being ignored in the daylight hours. A buzzer could be connected to each warning light, or a relay connected to each light with the contacts operating a buzzer—but a much simpler approach would be to use guide or steering diodes and a single common buzzer.

Reference to the circuit depicted in Figure 2.19 will explain the common warning operation. This is a simplified circuit of a motor vehicle's warning system. The water switch is normally open and will close with an excessive rise in temperature. The oil switch is normally closed but opens when the oil pressure is at the correct level. Once the engine is running, a malfunction will cause either switch to close and light its respective lamp.

With both the switch contacts open, the lower side of the lamp has a positive potential and as the buzzer is also connected to the positive line no current flows. Now if either switch closes, the lower side of its respective lamp becomes negative, its diode is forward biased, and the buzzer sounds. The other lights are isolated by their diodes and so only the one lamp operates.

The charge light is connected to a point which is normally positive when the engine is running and which turns negative when the engine is stopped or when the alternator ceases to charge due to a malfunction. Once

again the buzzer will sound if the 'aux' point becomes negative and the charge lamp lights. The diodes function as guides and prevent current flowing except in their own circuits. This prevents the other lights from lighting and causing confusion. (It must be noted that when the ignition is switched on, the buzzer will sound until the engine is operating—signifying that the buzzer is working and is ready to indicate a malfunction.)

Power supply isolation (motor vehicle batteries) Motor vehicles fitted with powerful driving lights, especially those equipped for night rally competition, employ two batteries. One is used for the general standard equipment of the vehicle and the other for the extra driving lights. It is desirable that the normal vehicle battery and the driving light battery be charged from the vehicle's alternator, but if a straight connection is made between each battery and the alternator, the batteries are effectively in parallel and do not operate independently. By the use of *isolating* or guide diodes, independent operation may be obtained, and yet the two batteries may be charged simultaneously.

A simplified connection diagram for this system is illustrated in Figure 2.20. When the generated emf of the alternator is higher than the normal potential difference (tpd) of the battery (i.e. during the normal battery charging), current flows from the positive side of the alternator to each battery through its own isolating diode. The batteries will share the current, depending on their individual tpds and internal resistances.

As each battery is connected to its own load, the load draws current from the alternator during the time the alternator is charging. When the vehicle is stationary (i.e. not charging), each battery supplies its own load only. Current cannot flow between the batteries as it is blocked by the isolating diodes.

Even if the driving light battery was completely discharged, no current would flow from the other battery—the vehicle could be started and the driving light battery recharged. The same system can be used for a motorhome where the living-quarters battery can be made separate from the vehicle battery, and yet both may be charged from the vehicle's alternator.

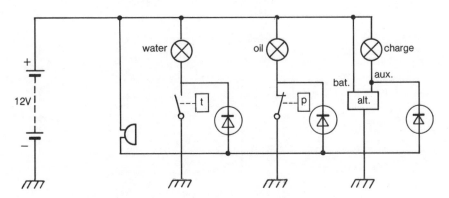

Fig. 2.19 *Buzzer added to warning lights in a car using guide diodes*

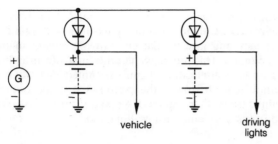

Fig. 2.20 *Vehicle electrical system and driving-lights supply, isolated from each other but charged from the same source*

Protection from self-induced emf If circuits with a relatively high inductance (for example coils of relays) are de-energised, the sudden collapse of magnetic flux will produce a high emf of self-induction which may damage the insulation of the coil. Additionally, the high emf may be impressed across other components in the circuit, causing considerable damage.

A simple method of eliminating this problem is to employ a diode, connecting it permanently across the coil. Reference to Figure 2.21 will show that the diode is normally reverse biased. When the supply to the coil is interrupted, the self-induced emf will tend to keep the current flowing in the original direction (Lenz's Law). This means that the diode is now forward biased and is a short circuit for the current produced by the emf of self-induction. This short circuit reduces the voltage of self-induction to a very low value (approximately 0.7 volt). Diodes used in this fashion are often referred to as *free-wheeling* or *flywheel* diodes.

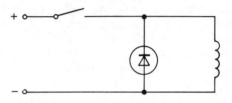

Fig. 2.21 *Discharge diode fitted across an inductive circuit*

Unit 2 **SUMMARY**

- A diode only allows current to flow through it in one direction.
- A PN diode is made from a layer of P-type and N-type semiconductor fused together.
- Between the two layers in a PN diode is a depletion layer which has an inbuilt potential that must be overcome before conduction can take place. This inbuilt potential is approximately 0.3 volt for germanium and 0.7 volt for silicon.
- Power diodes are invariably silicon diodes.
- When a potential is applied so that a diode conducts, we say that it is forward biased. If the polarity is reversed, and no conduction takes place, we say that it is reverse biased.
- An increase in temperature in a PN diode results in an increase in leakage current and a reduction in forward voltage drop.

- A zener diode will conduct when reverse biased only when the applied potential reaches a certain predetermined point. This point is called the zener point and determines the zener voltage.
- LEDs are special diodes which emit light when carrying current.
- A photodiode operates in the reverse biased condition, with leakage current dependent on the intensity of light falling on the junction.
- Diodes can be used in circuits so that currents will only flow in particular directions. When used in this manner they are referred to as guide or steering diodes. Other names used are isolation diodes, discharge diodes, flywheel diodes or free-wheeling diodes. The diode used is only a normal type and is given these names to indicate its usage in the particular circuit.

Rectifiers 1— single-phase

3.1 Rectification

The word 'rectify' literally means to 'make right'—and with reference to an electrical circuit, it might be assumed that 'to rectify' would be to make right something that was wrong in the first place. This is, of course, erroneous. (The whole idea of 'right 'and 'wrong' in electrical circuits is a carryover from the sometimes bitter rivalry between the interests promoting the distribution of electrical energy by either direct current or alternating current in the early days of electrical development at the end of the nineteenth century.)

Direct current, as well as being used for lighting, heating and motive power, can also be used for battery charging and electrochemical processes (electroplating, electrolytic refining and the like). Alternating current, because of its ease of distribution, as well as being able to use the simple induction motor, is now universal— however, it cannot be used for battery charging and electrochemical processes unless it is changed to direct current.

The proponents of direct current distribution gleefully pointed out that the rival alternating current had to be 'made right', or changed to direct current before it could be used for the above-mentioned purposes: the term *rectify* thus gaining use in the electrical sense.

Although rotating machines can be used to change ac to dc we usually call these *converters* and reserve the term *rectifier* for electronic devices that do the same job. The process of changing alternating current to unidirectional or direct current is called *rectification*. In this book we are only concerned with electronic rectifiers.

Although we will be examining a number of different rectifier circuits, in this unit and the next one we use the normal shortened rectifier symbol in many circuit representations (especially single line diagrams). This symbol is illustrated in Figure 3.1, which shows an incoming ac supply connected to the ac terminals of the device, emerging from the dc terminals as unidirectional or direct current. (If the arrow in the symbol was reversed, it would indicate that the device was an *inverter* which changes dc to ac.) On the symbol the sine wave represents the ac terminals and the small stroke the dc terminals.

Although the distribution of electrical energy is universally ac, there are many areas in which dc is ideal. One such instance is electric traction. For instance, in dc mainline traction in electric trains, the alternating supply is rectified to dc in sub-stations along the route, usually at 1500 volts, and applied to the overhead wiring and the track. Direct current is used because the dc series motor is supreme in traction work. Another, more modern system is to supply the overhead wiring with 25 kilovolts ac, and use transformers and rectifiers onboard the train to supply direct current to the motors. Diesel-electric locomotives generate ac which is rectified to dc to supply their traction motors.

Aluminium smelters, too, must use dc in the electrolytic refining process, and the incoming ac supply is rectified to dc by many hundreds of silicon rectifiers, supplying the extremely high current required. Such a smelter may have a power demand of 600 megawatts: a dc current of 180 000 amperes could flow through the electrolytic cells.

The battery in a car must similarly be charged with dc and the alternator, which supplies three-phase ac, must have its output rectified to dc by silicon diodes usually built into the alternator itself.

3.2 The rectifying action

A diode will only pass current in one direction, so consider just what will occur when an alternating current is applied to a diode. When the diode is forward biased, say during the positive half of the alternating current wave, current will flow through the diode with the same waveform as the original alternating current. When, however, the alternating current wave passes through zero and into the negative half-cycle, the diode is reverse biased, and no conduction will take place.

The result of this process is that only the current of each *positive* half wave flows. Such current, of course, only flows in pulses, but the important thing is that it *only flows in the one direction*. It can thus be said to be unidirectional, or direct current. (This output can be *smoothed* by the use of filters, which will be discussed in unit 5 of this book.) The diagrams in Figure 3.4 detail the result. Note that if the diode was reversed, only *negative* half cycles would pass, and the output polarity would be reversed.

The diodes used for rectification (and indeed for all power work) are invariably silicon, which are the most efficient ones that have been developed and can operate with an efficiency of up to 99.8 per cent.

3.3 Rectifier diode polarities

Reference to Figure 3.4 will indicate that the *cathode* of the diode is positive in rectifier service. This appears to be a contradiction of what was said in section 2.1, but a little thought will show that it is actually quite logical.

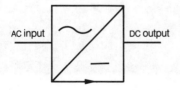

Fig. 3.1 *An abbreviated symbol for a rectifier circuit*

AC input   DC output

Fig. 3.2 *Electric traction must use dc. This train is supplied by the overhead wiring from rectified 1500 volts dc* STATE RAIL AUTHORITY OF NSW

Fig. 3.3(a) *The aluminium and smelting industry uses huge quantities of rectified dc energy* TOMAGO ALUMINIUM

Fig. 3.3(b) *Banks of diodes used to rectify the incoming ac to dc for aluminium smelting. Equipment by Siemens Ltd* *TOMAGO ALUMINIUM*

The diode used as a rectifier may be considered to be a switch: 'on' when forward biased and 'off' when reverse biased. Consider the diagram in Figure 3.5. The switch above each diode represents the conduction conditions of the diode. During the *positive* half-cycle, the diode is forward biased and conducts. Neglecting the *very small* forward voltage drop across the diode, the now positive side of the supply is connected directly to the top side of the load. The total voltage drop of the supply appears across the load with the *positive* end (the top end) connected to the diode cathode.

During the *negative* half-cycle, the diode is reverse biased. This is indicated by the open switch, and obviously no current is flowing in this condition. Also, as there is no voltage drop across the load resistor, the now positive side of the supply appears at the diode cathode.

Thus in rectifier service, under all conditions, a positive potential appears at the cathode of the diode. It is for this reason that on some diodes the cathode terminal is indicated by a red spot, indicating that this is the positive terminal in rectifier service.

If a moving-coil voltmeter (or one which acts similar to a moving coil type) is connected across the diode, it will read zero during each positive half cycle (again neglecting the very small forward voltage drop). During each negative half cycle, the meter will read the full voltage of the supply which appears across the reverse biased diode. At normal power frequencies the voltmeter will read the average voltage of the full cycle, which will be 0.45 of the ac rms voltage.

When drawing rectifier circuits it is always best to consider that current flows in the direction of the cathode, so the end of the load which connects to the cathode must be positive.

In the remainder of this unit and in the next one, the various common rectifier configurations will be examined, but to understand these circuits one must completely understand the fundamentals just discussed.

3.4 Single-phase half-wave rectifier

The single-phase half-wave, as shown in Figure 3.6 is the simplest type of rectifier circuit. Whenever a diode is inserted in series with a load and connected to an ac supply, the result is single-phase half-wave rectification. Obviously, the output voltage is somewhat less than the rms voltage of the ac supply. If measured with a moving iron instrument (or an instrument which reads rms values), it would be found that the rectified supply voltage is 0.5 of the ac rms voltage—or, to put it another way, it would be 0.3535 of the maximum value of the supply ac wave. The current that this voltage produces in a load (provided that it is non-inductive) would have the same heating effect as if it was produced by steady direct current.

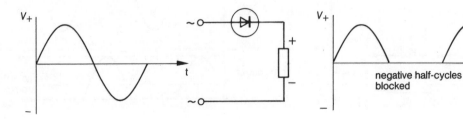

Fig. 3.4 *The rectifying action of a diode connected to an alternating supply*

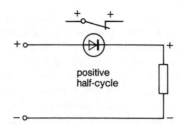

Fig. 3.5 *Polarities across rectifier diodes*

As stated previously, we must consider the unidirectional current from this circuit to be dc. For this reason it is usual to employ moving-coil instruments, which indicate the *average* value in rectified dc circuits. (Digital meters will also indicate the average value.)

In a steady dc circuit (as from a battery), the moving coil or equivalent instrument will measure the average value which is identical (in this case) with the *effective* (or rms) value of voltage or current. However, when connected in a *pulsating* supply circuit, the instruments will indicate an average value somewhat less than the effective value.

As rectified supplies are primarily used for motive power or electrochemical purposes (battery charging, electroplating, electrolytic refining and such) it is only the average value that has any real significance. If the heating effect of the current is to be considered, then it must be remembered that the slightly higher effective (rms) value must be applied. Despite this, it is generally accepted that all measurements in rectified supplies are made with moving coil instruments or instruments which indicate average values.

In rectification, the ratio of the output average value to the input rms value (or, as is usually expressed V_{DC} to V_{AC}) is usually quoted. In the case of the single-phase half-wave rectifier, the ratio is:

$$V_{DC} = 0.45 \ V_{AC} \qquad (3.1)$$

The above equation neglects the very small voltage drop across the diode (approximately 0.7 volt). If the rectified voltage of the circuit is high, say 200 volts, this small voltage drop is less than the order of accuracy

of most voltmeters and may be completely ignored. If, however, the rectified voltage was, say 20 volts, the difference would be nearly 10 per cent, which is significant. And if the rectified voltage was very low, say 2 volts, the difference would be 65 per cent—which cannot be accepted. The student must always note the rectified voltage of the circuit when considering whether or not to include the diode voltage drop in calculations.

Another point to be remembered is that if the load across the output of a rectifier includes a capacitor, the load voltage will be entirely different. If there is a capacitor alone as the load, the output will be the peak voltage of the wave, or V_{MAX}.

When the V_{DC} output voltage is applied to a load, the average current that flows in the load is simply determined by Ohm's Law, using the equation:

$$I_L = \frac{V_L}{R_L} \qquad (3.2)$$

where I_L is load current in amperes,
      V_L is the output voltage applied to the load
           in volts,
      R_L is the resistance of the load in ohms.

The waveform of the output consists of a series of pulses. These arrive at the same frequency as the ac supply as they are each part of a cycle. This is termed the *ripple frequency* and in this case:

$$f_R = f \qquad (3.3)$$

where f_R is the ripple frequency,
      f is the supply frequency.

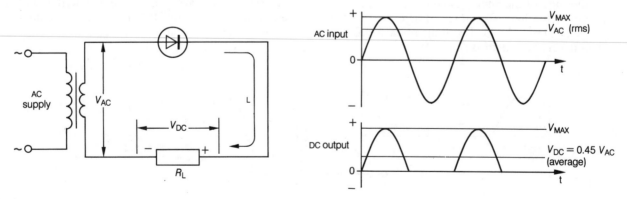

Fig. 3.6 *Single-phase half-wave rectifier circuit with input and output waveforms*

The periodic variation of the output between maximum and minimum values is termed the dc *ripple voltage*. In this case the dc ripple voltage is the complete half-wave. This may be expressed in the form:

$$V_{ripple} = V_{MAX} = \sqrt{2}\ V_{AC} \qquad (3.4)$$

where V_{ripple} is the dc ripple voltage,
V_{MAX} is the peak value of the wave,
V_{AC} is the ac input to the rectifier.

When the diode is not conducting on the negative half cycle, the full supply voltage appears across it. For this reason the maximum peak inverse voltage, or PIV_{MAX}, of the particular diode in use must be at least equal to (and in practice be much higher, for a margin of safety) the V_{MAX} of the rectified output. So that the correct diode may be selected for a circuit, it is usual to quote the maximum potential that could normally be expected across the diode in a rectifier circuit. This is usually called the peak inverse voltage (PIV) or peak reverse voltage (PRV) of the particular circuit. For this circuit this may be expressed as:

$$PIV = V_{MAX} = \sqrt{2}\ V_{AC} \qquad (3.5)$$

When selecting a diode to be used in a circuit, not only the PIV of the circuit but also any other voltage present in the circuit, must be taken into consideration. For example if the above circuit was used to charge a battery, on the non-conducting part of the cycle the battery voltage is effectively in series with V_{MAX} and so the PIV is V_{MAX} *plus* the emf of the battery V_{BAT}. A further consideration is when the load connected to the rectifier is a capacitor. In this case the PIV is doubled as the capacitor is fully charged to V_{MAX} and this is effectively in series with the peak value of the wave when on the non-conducting half cycle.

Although the simplest rectifier, the single-phase half-wave rectifier circuit is the least efficient and considered a poor choice except for very low power applications. One severe disadvantage with the circuit is that it causes dc to flow in the mains. If the current was high this could cause great problems with metering, and when the M.E.N. system of earthing is used, could cause electrolytic corrosion in earth electrode and water supply connections. This does not mean that the circuit is never used, but discretion is always employed before it is considered.

Example 3.1

A single-phase half-wave rectifier circuit is shown in Figure 3.7. Considering the circuit, determine the following:

(a) load voltage;
(b) load current;
(c) the PIV across the diode;
(d) the ripple voltage;
(e) ripple frequency.

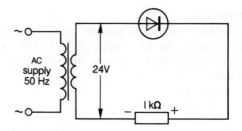

Fig. 3.7   *Circuit for Example 3.1*

$$\begin{aligned}
V_{AC} &= 24\ V \\
R_L &= 1\ k\Omega \\
f &= 50\ Hz \\
V_L &= ? \\
I_L &= ? \\
PIV &= ? \\
V_{ripple} &= ? \\
f_R &= ?
\end{aligned}$$

$$\begin{aligned}
V_L &= V_{DC} \\
&= 0.45\ V_{AC} \\
&= 0.45 \times 24 \\
&= 10.8\ V
\end{aligned}$$

Answer (a): The load voltage is 10.8 volts.

$$\begin{aligned}
I_L &= \frac{V_L}{R_L} \\
&= \frac{10.8}{1000} \\
&= 10.8 \times 10^{-3}\ A
\end{aligned}$$

Answer (b): The load current is 10.8 milliamperes.

$$\begin{aligned}
PIV &= \sqrt{2}\ V_{AC} \\
&= \sqrt{2} \times 24 \\
&= 33.94\ V
\end{aligned}$$

Answer (c): The PIV applied across the diode is 33.9 volts.

$$\begin{aligned}
V_{ripple} &= \sqrt{2}\ V_{AC} \\
&= \sqrt{2} \times 24 \\
&= 33.94\ V
\end{aligned}$$

Answer (d): The dc ripple voltage is 33.9 volts.

$$\begin{aligned}
f_R &= f \\
&= 50\ Hz
\end{aligned}$$

Answer (e): The ripple frequency is 50 hertz.

Example 3.2

A single-phase half-wave rectifier is used to charge a 50 volt battery. The ac voltage input to the circuit is 150 volts. What would be the peak inverse voltage applied to the diode?

V_{AC} = 150 V

V_{BAT} = 50 V

PIV = ?

$$V_{MAX} = \sqrt{2}\ V_{AC}$$
$$= \sqrt{2} \times 150$$
$$= 212\ V$$
$$PIV = V_{MAX} + V_{BAT}$$
$$= 212 + 50$$
$$= 262\ V$$

Answer: The PIV applied across the diode is 262 volts.

Example 3.3

A single-phase half-wave rectifier circuit has an ac input voltage of 32 volts. A capacitor is connected across the load terminals. What is the load voltage and what is the PIV across the diode?

V_{AC} = 32 V

V_{L} = ?

The capacitor will charge up to the peak value of the half wave and as there is no discharge path for the capacitor, this will be the load voltage.

$$V_{MAX} = \sqrt{2}\ V_{AC}$$
$$= \sqrt{2} + 32$$
$$= 45.25\ V$$

Answer: The load voltage is 45.3 volts.

As the capacitor is charged up to V_{MAX}, on the positive half-cycle, this voltage is in series with the, now, reverse peak voltage of the wave, so:

$$PIV = V_{MAX} + V_{MAX}\ (capacitor)$$
$$= 45.25 + 45.25$$
$$= 90.5\ V$$

Answer: The peak inverse voltage across the diode is 90.5 volts.

3.5 Single-phase full-wave centre-tap rectifier

As the single-phase half-wave rectifier is considered a poor choice, we will now examine a method of utilising each half of the ac wave to obtain full-wave rectification. If a second winding is added to the transformer in Figure 3.6 we could have a circuit in which the two windings could be connected together as in Figure 3.8. With the addition of another diode it can be seen that with reference to the joining of the two windings, or as it is usually termed, the *centre-tap*, each diode will have an opposite instantaneous polarity. Another way of saying this is that

each outer side of the winding is 180^0 out of phase, with reference to the centre-tap.

In the input curve of Figure 3.8, one waveform is shown solid and the other is shown dashed. These represent the out-of-phase voltages on each side of the centre-tap. When point *e* in the diagram is positive with respect to point *o* (the centre-tap), diode D1 is forward biased and current will flow through it to the positive side of the load and to the centre-tap. During this time, point *f* is negative and diode D2 is reverse biased.

On the next half-cycle the reverse takes place and point *f* is positive. Diode D2 now conducts and current flows through it to the positive side of the load and to the centre-tap. This, of course, continues and the current flowing through R_L consists of half sine-waves *but* all of the same polarity. This means that we have now utilised both halves of the ac sine wave, hence the name full-wave. This is a much better system than half-wave, not only for the line and transformer (because of the ac wave distortion and greatly reduced efficiency) but also for the load, as there are now no blank periods in the output voltage and current.

Note that except for the unidirectional nature of the output, the output is, neglecting diode voltage drop, the same as the input. This would indicate that the rms value of the output is identical with the input. However, as it is conventional to use average values for the dc output of rectifiers (for reasons already mentioned) we can say:

$$V_{DC} = 0.9\ V_{AC} \qquad (3.6)$$

where V_{DC} is rectifier output voltage,
V_{AC} is ac input voltage.

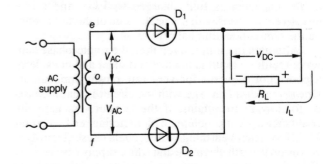

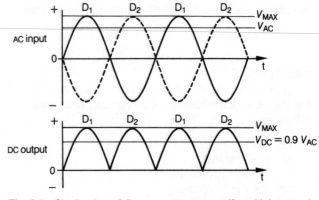

Fig. 3.8   *Single-phase full-wave centre-tap rectifier with input and output waveforms*

Also, in the circuit in Figure 3.8 V_{DC} equals V_L. Once more we must note that diode voltage drop has been neglected and if a capacitor is connected across the load the value of V_L can vary greatly.

The average value of the current flowing in the load, R_L, may be determined from the same equation (3.2) as before:

$$I_L = \frac{V_L}{R_L}$$

Now as we are using both halves of the wave it must follow that the ripple frequency is twice that of the half-wave circuit, so:

$$f_R = 2f \qquad (3.7)$$

where f_R is the ripple frequency,
   f is the supply frequency.

Although we are using the full wave, the dc ripple voltage is still the same, as it is measured from where each half-wave intersects, the zero line, to the maximum. This is the same as equation 3.4:

$$V_{ripple} = \sqrt{2}\, V_{AC}$$

As we now have two transformer windings effectively in series, as far as the diodes are concerned, the PIV across each diode must be twice the V_{MAX} of each winding. This is written as:

$$PIV = 2\sqrt{2}\, V_{AC} \qquad (3.8)$$

This is the greatest voltage that can appear across each diode and the connection of a battery or capacitor to the load terminals will have no effect on the diode PIV.

The single-phase full-wave centre-tap rectifier is a very useful circuit. Its main advantage lies in the fact that only two diodes are needed for full-wave output. This could be important for two reasons: first, the diode voltage drop is at a minimum (if this is to be considered) and second, in high current ratings diodes can be very expensive. Against this a centre-tap transformer must be provided and this means not only the expense of the double secondary winding but a transformer efficiency of only about 64 per cent as only one half of the secondary is used for each half-cycle.

As two diodes are sharing the current, the current rating of each diode need only be one half the average output current. For example, if the average output current was to be 24 amperes, only 12 ampere diodes need be selected for the circuit.

Example 3.4

For the rectifier circuit in Figure 3.9, determine the following:

   (a) load voltage;
   (b) load current;
   (c) diode PIV;
   (d) ripple frequency;
   (e) dc ripple voltage.

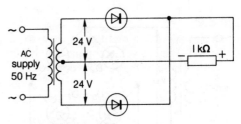

Fig. 3.9   *Circuit for Example 3.4*

$$\begin{aligned} V_{AC} &= 24\ V \\ R_L &= 1\ k\Omega \\ f &= 50\ Hz \\ V_L &= ? \\ I_L &= ? \\ PIV &= ? \\ f_R &= ? \\ V_{ripple} &= ? \end{aligned}$$

$$\begin{aligned} V_L &= V_{DC} \\ &= 0.9\ V_{AC} \\ &= 0.9 \times 2 \\ &= 16.2\ V \end{aligned}$$

Answer (a): The load voltage is 16.2 volts.

$$\begin{aligned} I_L &= \frac{V_L}{R_L} \\ &= \frac{16.2}{1000} \\ &= 16.2 \times 10^{-3}\ A \end{aligned}$$

Answer (b): The load current is 16.2 milliamperes.

$$\begin{aligned} PIV &= 2\ \sqrt{2}\ V_{AC} \\ &= 2 \times \sqrt{2} \times 24 \\ &= 67.88\ V \end{aligned}$$

Answer (c): The peak inverse voltage across the diodes is 67.9 volts.

$$\begin{aligned} f_R &= 2f \\ &= 2 \times 50 \\ &= 100 \end{aligned}$$

Answer (d): The ripple frequency is 100 hertz.

$$\begin{aligned} V_{ripple} &= \sqrt{2}\ V_{AC} \\ &= \sqrt{2} \times 24 \\ &= 33.94\ V \end{aligned}$$

Answer (e): The dc ripple voltage is 33.9 volts.

3.6 Single-phase full-wave bridge rectifier

By employing four diodes it is possible to eliminate the double secondary transformer and still obtain full-wave rectification. Figure 3.10 shows the full-wave *bridge circuit*, so named because of its resemblance to the familiar bridge measuring circuits.

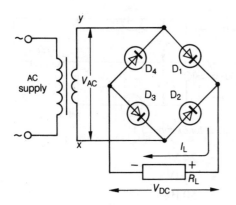
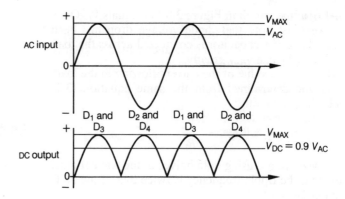

Fig. 3.10 *Single-phase full-wave bridge rectifier circuit with input and output waveforms*

This circuit is able to produce full-wave rectification by using the two extra diodes as *guide diodes*. Let us study the current paths in the bridge that produce this result.

Assume that point x is positive and point y is negative. Diode D_2 is forward biased and diode D_1 reverse biased. Current therefore flows from point x via diode D_2 to the positive terminal, through the load, and, as the potential at diode D_3 on the cathode side is higher than the anode side (reverse biased), current can only flow through diode D_4 to y.

Now, assume point y is positive and point x is negative. Current is blocked at diode D_4 but flows through diode D_1. Diode D_2 then blocks, but current flows through the load from positive terminal to negative, through diode D_3 (D_4 is reverse biased) to point x.

It can be seen that in each case the direction of the current through the load is the same. The output is similar to the centre-tap circuit, so the output voltage equation is equation 3.6:

$$V_{DC} = 0.9\, V_{AC}$$

Again the load current is calculated by the Ohm's Law equation 3.2:

$$I_L = \frac{V_L}{R_L}$$

The ripple frequency is exactly the same as the centre-tap circuit, equation 3.7:

$$f_R = 2f$$

The dc ripple voltage is also identical with the centre-tap circuit, and indeed the half-wave circuit, equation 3.4:

$$V_{ripple} = \sqrt{2}\, V_{AC}$$

The diode peak inverse voltage is only the maximum of the one winding (not the double winding like the centre-tap circuit) and so we can use equation 3.5:

$$PIV = \sqrt{2}\, V_{AC}$$

As for the centre-tap circuit, the current flowing in the diodes is one half of the average current flowing in the load. The bridge circuit is the most used of the

Fig. 3.11 *Two typical small bridge rectifiers*

single-phase rectifier circuits and many manufacturers make up bridge circuits in a single package (including the inbuilt four diodes). Figure 3.11 shows two rectifier bridges. Note the symbols for the ac input and dc output connections. Compare this with the abbreviated schematic symbol illustrated in Figure 3.1.

Figure 3.12 illustrates two other configurations of the single-phase bridge rectifier circuit. The configuration on the left is often used in preference to the one illustrated in Figure 3.10, as the layout is similar to the three-phase circuits to be covered in the next unit.

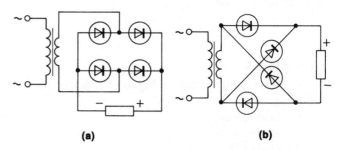

(a)   **(b)**

Fig. 3.12 *Two further configurations of the bridge rectifier circuit. Both diagrams are electrically identical*

Example 3.5

Assuming a single-phase bridge rectifier has an ac input voltage of 24 volts at 50 hertz, and a load resistance of 1 kilohm, determine the following:

(a) output voltage;
(b) load current;
(c) ripple frequency;
(d) dc ripple voltage;
(e) diode PIV.

$$V_{AC} = 24 \text{ V}$$
$$f = 50 \text{ Hz}$$
$$R_L = 1 \text{ k}\Omega$$
$$V_L = ?$$
$$I_L = ?$$
$$f_R = ?$$
$$V_{ripple} = ?$$
$$PIV = ?$$

$$V_{DC} = 0.9 \, V_{AC}$$
$$= 0.9 \times 24$$
$$= 21.6 \text{ V}$$

Answer (a): The average output voltage is 21.6 volts.

$$V_L = V_{DC}$$
$$I_L = \frac{V_L}{R_L}$$
$$= \frac{21.6}{1000}$$
$$= 21.6 \times 10^{-3} \text{ A}$$

Answer (b): The average load current is 21.6 milli-amperes.

$$f_R = 2f$$
$$= 2 \times 50$$
$$= 100 \text{ Hz}$$

Answer (c): The ripple frequency is 100 hertz.

$$V_{ripple} = \sqrt{2} \, V_{AC}$$
$$= \sqrt{2} \times 24$$
$$= 33.94 \text{ V}$$

Answer (d): The dc ripple voltage is 33.9 volts.

$$PIV = \sqrt{2} \, V_{AC}$$
$$= \sqrt{2} \times 2$$
$$= 33.94 \text{ V}$$

Answer (e): The diode peak inverse voltage is 33.9 volts.

Unit 3 **SUMMARY**

- Rectification is changing ac to dc with electronic rectifiers. Inversion is changing dc to ac with electronic inverters.
- All power rectifiers use silicon diodes.
- Direct current is essential for electric traction, battery charging and electrolytic refining.
- The simplest rectifier circuit consists of an ac supply and a diode in series with a load. This is called a single-phase half-wave rectifier.
- The cathode of a diode is always considered positive in rectifier service.
- The single-phase half-wave rectifier circuit is seldom used as it is inefficient, causes problems with the mains and has a high diode PIV.

- The single-phase full-wave centre-tap rectifier circuit must use a double winding centre-tapped secondary transformer.
- The most commonly used single-phase rectifier circuit is the bridge rectifier. It produces the same output as the centre-tapped circuit but has only one half the diode PIV.
- Capacitors connected across the load will change the load voltage of a rectifier.
- If the load of a rectifier is only a capacitor, the load voltage will equal the peak value of the input wave.

Unit 4

Rectifiers 2— three-phase

4.1 Single-phase versus multi-phase

In the single-phase rectifier circuits discussed in Unit 3 the most noticeable feature in the output waveforms was the high ripple. In the case of the half-wave circuit the ripple, as it stands, is intolerable—and possibly the only application would be off load battery charging. Quite often the ripple in the output is expressed as a ratio between the rms value of the ripple content and the average dc output. In the case of the half-wave circuit this can be shown to be 121 per cent.

The single-phase full-wave circuits are certainly very much better than the half-wave, but still the instantaneous voltage falls to zero each half cycle. The ripple voltage is equal to the maximum, or peak, value of the wave and the ripple is 48 per cent. In low power applications it is possible to remove this ripple with filters, but when considering higher power rectifiers (with say currents exceeding 10 amperes and voltages exceeding 200 volts) the filtering can become quite expensive and uneconomical.

For this reason multiphase rectifier circuits are always used for higher power outputs. Normally the higher the power, the more phases, and whereas three phases are most commonly used, it is possible to use six-, twelve- and twenty-four-phase circuits. In this book we will only refer to three-phase and look briefly at six-phase.

4.2 Three-phase half-wave rectifier

In a three-phase system, where each phase has an angular time difference of 120°, it becomes apparent that if each phase is utilised in a rectifier system there are three positive pulses in the same time as a complete cycle of one single-phase. (There are also, of course, three negative pulses, but in a half-wave system these are ignored.)

Referring to Figure 4.1 it can be seen that there is a diode in series with each phase of the input. To obtain a reference point to these three phases it must be apparent that a star connected transformer is *essential*, and the neutral (star point) becomes a rectifier terminal (the primary of the transformer, however, may be connected in either star or delta). It is usual to make the star point the negative terminal and so the three cathodes of the rectifiers are connected together to make the positive terminal. (If, however, the star point was to be made positive, the three diodes are simply reversed and the three anodes, connected together, would make the negative terminal.)

In Figure 4.1, which considers a phase rotation of A, B, C, the most positive phase, at any instant, will forward bias its own diode and reverse bias the diodes in the other two phases. In the input wave diagram only the wave for one phase is shown for clarity, but in the output diagram each positive pulse is lightly drawn with the actual output voltage waveform in heavy lines. Each diode will conduct for 120°. As each diode only conducts for one third of each cycle the current carried by each diode is only one third of the load current.

In this, as in all multiphase rectifiers, the average output voltage, or V_{DC}, is greater than the rms input, or V_{AC}. This value is:

$$V_{DC} = 1.17\ V_{AC} \qquad (4.1)$$

where V_{DC} is average output voltage,
V_{AC} is the ac input phase voltage.

In this case V_{AC} is the *phase* voltage of the ac input and in all rectifier calculations involving multiphase circuits V_{AC} is always quoted as the phase value and not *line* value.

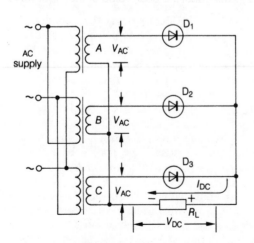
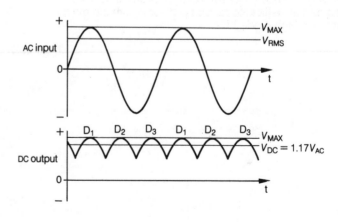

Fig. 4.1 *Three-phase half-wave rectifier with input and output waveforms*

The average load current, as in all rectifier circuits, is the Ohm's Law equation 3.2:

$$I_L = \frac{V_L}{R_L}$$

As stated before there are three pulses in periodic time so the ripple is:

$$f_R = 3f \qquad (4.2)$$

where f_R is the ripple frequency,
f is the supply frequency.

The diode PIV is the peak value of the *line* voltage but when referred to phase voltage (V_{AC}) it becomes $\sqrt{3}$ times $\sqrt{2}$ or:

$$PIV = 2.45\ V_{AC} \qquad (4.3)$$

where PIV is the peak inverse voltage applied to the diodes,
V_{AC} is the ac phase voltage of the input.

Instead of the ripple voltage being the full peak value of the wave it is now only one half peak value. This is a ripple on the dc output. For this circuit the ripple voltage is:

$$V_{ripple} = 0.354\ V_{AC} \qquad (4.4)$$

where V_{ripple} is the dc ripple voltage,
V_{AC} is the ac input phase voltage.

It can be readily seen that the ripple content of the three-phase half-wave circuit is much less than any single-phase circuit as at no time does the instantaneous voltage drop to zero. The ripple can be shown to be 18.3 per cent.

Example 4.1

For the simplified three-phase half-wave rectifier circuit shown in Figure 4.2, determine:

   (a) average output voltage;
   (b) average load current;
   (c) ripple frequency;
   (d) dc ripple voltage;
   (e) diode PIV.

$$V_{AC} = 24\ V$$
$$R = 1\ k\Omega$$
$$f = 50\ Hz$$
$$V_{DC} = 1.17\ V_{AC}$$
$$= 1.17 \times 24$$
$$= 28.08\ V$$

Answer (a): The average dc output voltage is 28.1 volts.

$$V_L = V_{DC}$$
$$I_L = \frac{V_L}{R_L}$$
$$= \frac{28.08}{1000}$$
$$= 28.08 \times 10^{-3}A$$

Answer (b): The average load current is 28.1 milli-amperes.

$$f_R = 3f$$
$$= 3 \times 30$$
$$= 150\ Hz$$

Answer (c): The ripple frequency is 150 hertz.

$$V_{ripple} = 0.354\ V_{AC}$$
$$= 0.354 \times 24$$
$$= 8.496\ V$$

Answer (d): The dc ripple voltage is 8.50 volts.

$$PIV = 2.45\ V_{AC}$$
$$= 2.45 \times 24$$
$$= 58.8\ V$$

Answer (e): The diode PIV is 58.8 volts.

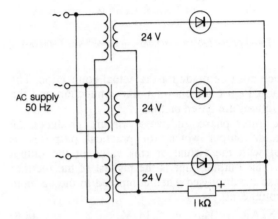

Fig. 4.2 *Circuit for Example 4.1*

4.3 Three-phase full-wave rectifier

In this circuit use is made of each half-cycle of each phase of a three-phase system. At any one time there are two diodes conducting and they conduct for 120°. Similar to the three-phase half-wave circuit each diode only carries one third the average load current. As there are now twice as many pulses in periodic time, as in the three-phase half-wave circuit, a much smoother output waveform is obtained.

The circuit does not require a transformer if the mains voltage is suitable; if a transformer is used (and usually for safety reasons it is) the secondary may be connected in either star or delta. The choice of a star or delta secondary will have a bearing on the ratio of V_{DC} to V_{AC}.

If the phase voltage is the *same* in both cases, the star connection will produce an output voltage which is $\sqrt{3}$ times the delta voltage. It must be noted then that when V_{AC} is quoted it refers to the phase voltage, and

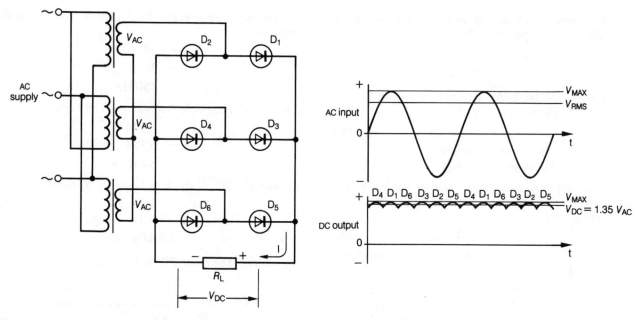

Fig. 4.3 *Three-phase full-wave rectifier circuit with input and output waveforms*

allowance must be made for the actual connection. The diagram in Figure 4.3 shows a star connection and the quoted figures are based on this.

The three-phase full-wave circuit produces an average dc output which, for practical purposes, is identical with the virtual or rms value of the output current. This output value is the highest of the rectifier circuits so far discussed when compared to the ac input phase voltage, as:

for STAR V_{DC} = 2.34 V_{AC} **(4.5)**
where V_{DC} is average output voltage,
$\quad V_{AC}$ is ac input phase voltage.

(*Remember* that if the connection was delta the average dc output voltage would be reduced by $\frac{1}{\sqrt{3}}$, i.e. reduced to 1.35 V_{AC}.)

The average dc load current once again is determined by using equation 3.2:

$$I_L = \frac{V_L}{R_L}$$

As this circuit uses each half cycle of a three-phase system it must follow that:

$$f_R = 6f \qquad \textbf{(4.6)}$$

where f_R is output voltage ripple frequency,
$\quad f$ is supply frequency.

The ripple voltage, as previously stated, is the smallest of all the rectifier circuits examined so far in this book. As the sine waves of each half-wave now intersect at angles equivalent to the sine of 60^0 the ripple voltage is:

$$V_{ripple} = 0.094 \ V_{AC} \qquad \textbf{(4.7)}$$

where V_{ripple} is the dc ripple voltage,
$\quad V_{AC}$ is the ac phase input voltage.

The peak inverse voltage that would be applied to the diodes is the peak value of the ac phase voltage (whether the connection is star or delta). This would be identical with the single-phase half-wave circuit, equation 3.5:

$$PIV = \sqrt{2} \ V_{AC}$$

For only the price of three extra diodes the three-phase full-wave rectifier circuit gives a much better result than does the half-wave. However, where cost is a factor (and the price of high-current diodes is not cheap), the three-phase half-wave circuit may be considered. The full-wave circuit has a ripple of only 4.2 per cent which, although not nearly as good as a dc generator, is often quite satisfactory. For higher powered battery chargers it would be the one most often used.

Example 4.2

Figure 4.4 shows the circuit of a three-phase full-wave rectifier. For the values given, determine:

(a) the average output voltage;
(b) the average load current;
(c) the ripple frequency;
(d) the dc ripple voltage;
(e) the diode peak inverse voltage.

$$V_{AC} = 24 \ V$$
$$R_L = 1 \ k\Omega$$
$$f = 50 \ Hz$$

$$\begin{aligned} V_{DC} &= 2.34 \ V_{AC} \\ &= 2.34 \times 24 \\ &= 56.16 \ V \end{aligned}$$

Answer (a): The average output voltage is 56.2 volts.

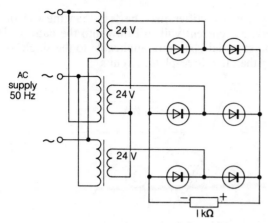

Fig. 4.4 *Circuit for Example 4.2*

$$V_{DC} = V_L$$

$$I_L = \frac{V_L}{R_L}$$

$$= \frac{56.2}{1000}$$

$$= 56.2 \times 10^{-3}A$$

Answer (b): The load current is 56.2 milliamperes.

$$f_R = 6f$$

$$= 6 \times 50$$

$$= 300 \text{ Hz}$$

Answer (c): The ripple frequency is 300 hertz.

$$V_{ripple} = 0.094 \; V_{AC}$$

$$= 0.094 \times 24$$

$$= 2.256 \text{ V}$$

Answer (d): The dc ripple voltage is 2.26 volts.

$$PIV = 2 \; V_{AC}$$

$$= 2 \times 24$$

$$= 33.94 \text{ V}$$

Answer (e): The diode peak inverse voltage is 33.9 volts.

Example 4.3

If the transformer secondary in the circuit of Figure 4.3 was reconnected in delta, what difference would there be in the answers obtained in Example 4.2?

$$V_{AC} = 24 \text{ V}$$

$$R_L = 1 \text{ k}\Omega$$

$$f = 50 \text{ Hz}$$

$$V_{DC} = 1.35 \; V_{AC}$$

$$= 1.35 \times 24$$

$$= 32.4 \text{ V}$$

Answer (a): Reconnected in delta the average output voltage is 32.4 volts.

$$V_{DC} = V_L$$

$$I_L = \frac{V_L}{R_L}$$

$$= \frac{32.4}{1000}$$

$$= 32.4 \times 10^{-3}A$$

Answer (b): The load current in delta is 32.4 milliamperes.

The reconnection in delta does not affect the remainder of the answers.

4.4 Six-phase rectifiers

It is simple to obtain six phases from a three-phase supply by providing two windings on each secondary of each phase of a transformer. Now if one of the twin secondaries on each phase is reversed (180° in relation to one another) they will provide three additional phases giving a phase difference of 60°. This can be seen in the diagram of Figure 4.5, where in the left side the windings have been joined as a common centre tap (star) and on the right side the secondaries only have been rearranged to show their phase relationship. These two configurations are electrically identical. This is a six-phase half-wave rectifier.

The six-phase half-wave rectifier has the same V_{DC} to V_{AC} ratio, ripple, dc ripple voltage and ripple frequency as the three-phase full-wave circuit. However, the diode PIV is higher at 2.45 V_{AC} (same as three-phase half-wave). Why the six-phase half-wave excels, is because the current carried by each diode is only one sixth of the average load current (half the rating of the three-phase full-wave), and the transformer efficiency is very much higher (95.1 per cent as against 63.6 per cent for the three-phase full-wave). The only disadvantage is having to provide the extra secondary winding on each phase.

(It would also be possible to arrange the six secondaries in a full-wave connection, requiring twelve diodes—and producing an even smaller ripple, and a dc output only just below the peak value of the wave.)

4.5 Rectifier reference tables

As a quick reference, Table 4.1 summarises figures for all the rectifier circuits discussed in this unit and the previous one, so the student may refer to them for calculation purposes.

4.6 Battery charging

When rectifier circuits are used in battery charging, as previously mentioned the charging current is the average current as measured on a moving-coil instrument or one that reads average values.

Another important point in battery charging is that current will flow through the battery as long as the *peak value* of the rectified waveform is greater than the tpd of the battery. At times the average value of the dc output voltage of a rectifier may be less than the tpd of the battery, but current will still flow into the battery. This, of course, would be more applicable to the single-phase rather than the three-phase circuits.

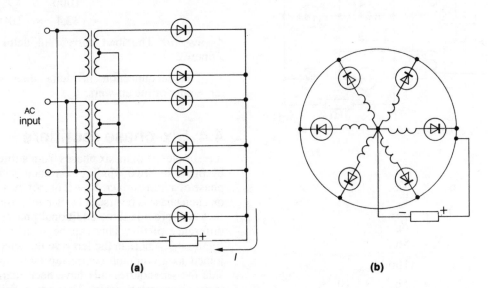

(a)   **(b)**

Fig. 4.5 *Six-phase half-wave star-connected rectifier circuit. (a) shows the complete circuit, and (b) the secondaries only, in their phase relationship.*

Table 4.1 *Rectifier reference*

Features	Single-phase half-wave	Centre-tapped full-wave	Single-phase bridge full-wave	Three-phase half-wave	Three-phase bridge (star)	Three-phase bridge (delta)	Six-phase star	Comments
Number of diodes	1	2	4	3	6	6	6	or paralleled groups of diodes
Diode PIV	$1.414\ V_{AC}$	$2.828\ V_{AC}$	$1.414\ V_{AC}$	$2.45\ V_{AC}$	$2.45\ V_{AC}$	$1.414\ V_{AC}$	$2.45\ V_{AC}$	Maximum voltage applied to diode when reverse biased
Average current per diode	I_{DC}	$\dfrac{I_{DC}}{2}$	$\dfrac{I_{DC}}{2}$	$\dfrac{I_{DC}}{3}$	$\dfrac{I_{DC}}{3}$	$\dfrac{I_{DC}}{3}$	$\dfrac{I_{DC}}{6}$	Assuming inductive load
Ripple frequency	f	$2f$	$2f$	$3f$	$6f$	$6f$	$6f$	f is supply frequency
Per cent ripple	121	48	48	18.3	4.2	4.2	4.2	$\dfrac{V_{rms}\ \text{ripple}}{V_{DC}}$
Output to input voltage ratio	0.45	0.9	0.9	1.17	2.34	1.35	1.35	Average, V_{DC} to rms phase voltage, V_{AC}, transformer secondaries
Transformer efficiency (per cent)	28.6	63.6	90	67.5	63.6	63.6	95.1	$\dfrac{\text{dc power output}}{\text{ac power input}}$

Note: V_{AC} refers to the *phase* voltage applied to the diodes. V_{DC} is the *average* output voltage

Fig. 4.6 *Three-phase full-wave battery charger rectifiers undergoing assembly*   POWER ELECTRONICS, A DIVISION OF WARMAN INTERNATIONAL LTD

Unit 4 SUMMARY

- When considering higher power outputs it is always preferable to use three-phase rectifiers.
- Three-phase half-wave rectifiers must always use a star connected transformer.
- In any rectifier, reversing all the diodes will reverse the polarity at the load.
- In any multiphase rectifier circuit, the average dc output voltage is always higher than the V_{AC} input phase voltage.
- In a three-phase full-wave rectifier circuit, the ratio dc average output voltage to ac input phase voltage (V_{DC} to V_{AC}), depends on whether the transformer secondaries are connected in star or delta.
- Six-phase half-wave rectifiers have an identical output to a three-phase full-wave circuit but each diode only carries one half the current and the transformer efficiency is much higher.
- When rectifiers are used for battery charging, current will flow into the battery as long as the peak value of each wave is less than the battery tpd.

Unit 5

Filters

5.1 The nature of filters

In electronic circuits, a filter can be described as any network which discriminates between different frequencies. This means a filter will exhibit a substantially constant low impedance over any desired range of frequencies and a high impedance for all other frequencies. (It must be noted that additionally there are *active* filters which contain amplifiers—but these are outside the scope of this book.)

As could be expected, in simple filter circuits there is no sharp transition between 'blocked' and 'passed' frequencies—but usually a transition range, with either blocked or passed frequencies gradually changing from fully passed to blocked as the frequencies increase or decrease.

The names given to filter circuits depend on their function: whether they block low frequencies and pass all others; whether they block all high frequencies and pass all others; whether they block middle frequencies and pass both high and low frequencies; or whether they pass all middle frequencies and block both high and low.

5.2 Filter types

As mentioned above, filters are classified according to the function they perform. The first one we will briefly examine is the *low pass filter*. Figure 5.1 shows the abbreviated symbol and a typical frequency response curve for this type.

This symbol is in a form similar to the rectifier symbol of Figure 3.1. The three sine waves inside the outline represent (from the top): high range, medium range and low range of frequencies. It can be seen that the two upper sine waves have a small line across them indicating that high and medium frequencies are blocked. The lower sine wave is not marked, indicating that it is passed. From this, of course, it gets its name.

The graph represents output voltage plotted against frequency. It is considered that all incoming frequencies have the same input voltage, and depending on the action of the filter the output voltage is affected. This filter, being a low pass type, does not affect the output voltage of the low frequencies—but at a certain point the output

voltage starts to drop until reaching an extremely low value (low enough to be considered zero). (The cut-off frequency, for reference only, is when the output voltage has dropped by about 30 per cent.)

The circuit of a very simple low pass filter is given in Figure 5.2. The capacitor has a high reactance to low frequencies and a low reactance to high frequencies. It can be seen that as the frequency gets higher the current is shunted across the load through the capacitor and its output voltage drops. We will discuss this type of filter later in this unit with regard to removing ripple from rectified supplies.

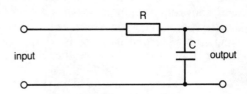

Fig. 5.2 *The circuit diagram of a simple low pass filter*

The high pass filter, Figure 5.3, has the same style of symbol except that the lower two sine waves have a small line across them, indicating they are stopped and only the higher frequencies are passed. The graph accompanying the symbol shows that only the higher frequencies are passed without attenuation. A simple high pass filter circuit is illustrated in Figure 5.4.

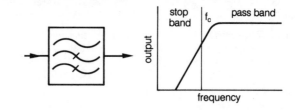

Fig. 5.3 *The symbol and performance graph of a high pass filter*

Band pass filters only allow a certain range of frequencies through—say from 300 hertz to 3000 hertz. The symbol and response graph of this type of filter is shown in Figure 5.5(a). The symbol clearly shows that upper and lower frequencies are blocked and the middle frequencies are passed. A simple circuit is shown in 5.5(b).

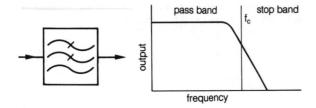

Fig. 5.1 *The symbol and performance graph of a low pass filter*

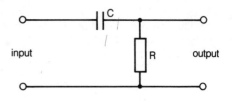

Fig. 5.4 *The circuit diagram of a simple high pass filter*

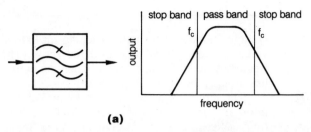

Fig. 5.5 *(a) The symbol and graph of a band pass filter (b) circuit diagram*

The opposite kind of filter to the band pass type is the *band stop* or *notch* filter. This lets through all frequencies above and below a certain range. Thus on the symbol the centre sine wave is marked off. The symbol of the band stop filter and its response graph are seen in Figure 5.6(a). A simple circuit is shown in 5.6(b).

5.3 Power supply filters

In both Units 3 and 4 mention was made of the ripple content of rectified supplies and how this ripple may be removed with filters. All rectified supplies have some ripple: from the very worst case (the single-phase half-wave circuit) to the twenty-four-phase full-wave rectifier (where it is barely discernible).

This ripple may or may not be embarrassing, depending on the purpose one has in mind for the equipment. In battery charging, where the batteries are off load, ripple is completely unimportant. Where the batteries are on load and being charged at the same time, a small ripple is unimportant as the batteries tend to absorb the ripple and leave the load free of ripple. This can be seen in telephone exchanges, where the batteries are continually on charge while delivering current for the telephone system. In these systems the battery actually 'floats' and the rectifier supplies the current to the system, the battery absorbing the small ripple.

In sound equipment, however, any ripple is quite unacceptable. Moreover, if sound equipment is operating from a single-phase supply—and this is almost universally the case—great effort must be made to reduce the ripple

to negligible proportions. Of course, only full-wave rectification would be used (except in very rare circumstances) and most filters are designed around this.

When power supplies are used for industrial processes ripple may interfere with performance. Motors may exhibit increased heat losses, and relays designed for dc may 'chatter' if the ripple amplitude is pronounced.

The output waveforms in Units 3 and 4 show that the smaller the amplitude of the ripple the greater the ripple frequency. For small current demand from rectifier supplies, ripple can quite easily be removed, even from single-phase half-wave circuits. When current demand is quite large, say in the order of 100 amperes, ripple is quite difficult and expensive to remove—so in this case one resorts to a circuit supplying a small amplitude and high frequency of ripple, e.g. a three-phase full-wave rectifier at least. When current demand is very high, say in the order of thousands of amperes, then possibly a twenty-four-phase full-wave circuit could be employed.

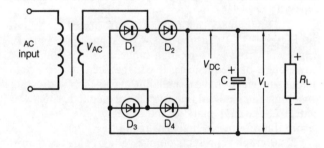

Fig. 5.7 *Schematic diagram of a capacitor filter power supply*

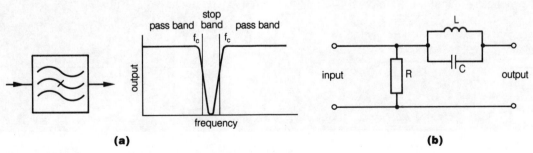

Fig. 5.6 *(a) The symbol and graph of a band stop filter (b) circuit diagram*

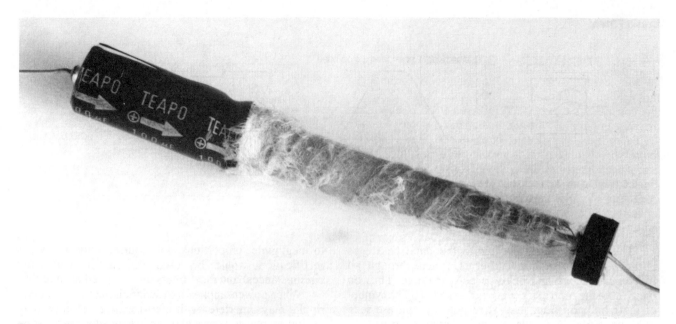

Fig. 5.8 *The results of connecting an electrolytic capacitor with incorrect polarity. The capacitor has exploded!*

Fig. 5.9 *Electrolytic capacitors (left) in the power supply of an amplifier constructed by the author*

5.4 Filter circuits—capacitors

The simplest filter circuit, and the one most commonly employed on low current demands, is the *capacitor filter*. This consists (as can be seen in Fig. 5.7 on p. 35) of a capacitor in parallel with the load. This is a very simple low pass filter—usually the capacitor is the polarised electrolytic type because a very high capacitance may be produced in a small space.

Because the capacitor is polarised, persons connecting them in the circuit must make certain that the correct polarity is observed, otherwise the capacitor may literally explode. (See Fig. 5.8.) Polarity markings on capacitors can vary between manufacturers but the two most common markings are a plus and minus sign to indicate positive and negative terminals, or a series of stylised arrows to indicate conventional current direction. Current *leaves* from the negative terminal. Figure 5.9 shows the polarity markings on electrolytic capacitors.

In the circuit shown in Figure 5.10, representing a single-phase full-wave rectifier, the current has been divided into three parts: the diode current; the capacitor current; and the load current. The waveform of these three currents is different, as they each have a different function. Figure 5.11 shows the rectifier and filter output curves, the input ac voltage curve and, below this, the load voltage curves. Two curves are shown depicting two situations: no load and load.

At no load, the output voltage may be given by the equation:

$$V_L \text{ (no load)} = \sqrt{2} \; V_{AC} \qquad (5.1)$$

where V_L (no load) is the output voltage at no load,
V_{AC} is the ac input voltage.

The other load voltage curve represents a typical load condition. The average value of this load current may be obtained from the approximate equation:

$$V_L \approx V_M \left[1 - \frac{T}{4 R_L C} \right] \qquad (5.2)$$

where V_L is load voltage,
T is periodic time in seconds,
R_L is load resistance in ohms,
C is capacitance in farads.

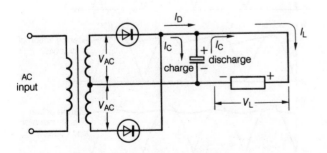

Fig. 5.10 *A simple single-phase full-wave power supply showing current paths*

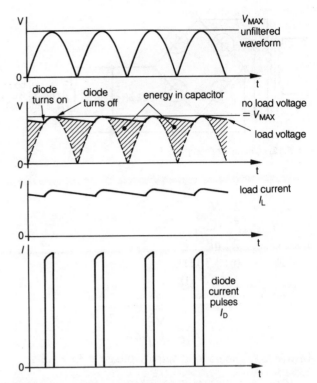

Fig. 5.11 *Output waveforms for capacitor filtered power supply*

The load current is in direct proportion to the load voltage and so follows the same waveform. The diode current, and transformer current, however, are entirely different (lower curves in Figure 5.11). It will be noted that the capacitor charges up to the peak value of the wave and as the current flows into the load the load voltage falls to a point on the next half-wave, where again it rises to the peak value as the capacitor recharges. Current only flows through the diodes during this capacitor recharging period. Since in this period the diode must supply energy for the load over the full half-cycle, it follows that the current will be relatively high: it is possible for the diode peak current to be six times the average dc output current.

If necessary this current may be limited by a series resistor, but usually if the diode has a high enough rating the series impedance of the transformer is quite sufficient. This current pulse has a high rms-to-average ratio due to its shape, and so heating in the transformer winding may be greater than what average current may indicate. The capacitor current (not shown on the graph) is the difference between the load current waveform and the diode current pulse.

Example 5.1

For the circuit of Figure 5.12 determine the following:

(a) no load dc output voltage;
(b) average dc load voltage when the load is 3.3 kilohms;
(c) average dc load voltage when the load is 1 kilohm.

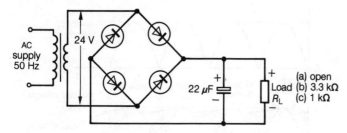

Fig. 5.12 *Circuit for Example 5.1*

$$V_{AC} = 24 \text{ V}$$
$$f = 50 \text{ Hz}$$
$$C = 22 \text{ } \mu\text{F}$$
$$R_L = \text{(b) 3.3 k}\Omega$$
$$\text{(c) 1 k}\Omega$$

(a)
$$V_L \text{ (no load)} = \sqrt{2} \text{ } V_{AC}$$
$$= \sqrt{2} \times 24$$
$$= 33.94 \text{ V}$$

Answer (a): The no load output voltage is 33.9 volts.

(b)
$$V_M = \sqrt{2} \text{ } V_{AC}$$
$$= \sqrt{2} \times 20$$
$$= 33.94 \text{ V}$$

$$T = \frac{1}{f}$$
$$= \frac{1}{50}$$
$$= 20 \times 10^{-3} \text{ s}$$

$$V_L \approx V_M \left[1 - \frac{T}{4 R_L C} \right]$$

$$\approx 33.94 \left[1 - \frac{20 \times 10^{-3}}{4 \times 3300 \times 22 \times 10^{-6}} \right]$$

$$\approx 33.94 \text{ } (1 - 0.069)$$
$$\approx 33.94 \times 0.931$$
$$\approx 31.61 \text{ V}$$

Answer (b): The load voltage with a 3.3 kilohm resistor is 31.6 volts.

(c)
$$V_L \approx V_M \left[1 - \frac{T}{4 R_L C} \right]$$

$$\approx 33.94 \left[1 - \frac{20 \times 10^{-3}}{4 \times 1000 \times 22 \times 10^{-6}} \right]$$

$$\approx 33.94 \text{ } (1 - 0.23)$$
$$\approx 33.94 \times 0.77$$
$$\approx 26.23 \text{ V}$$

Answer (c): The load voltage with a 1 kilohm resistor is 26.2 volts.

It can be seen that with higher load current (less load resistance) the average dc load voltage falls. Referring back to the centre graph in Figure 5.10, these conditions may be readily seen. Although the ripple is very greatly reduced as compared to an unfiltered supply, there is still some ripple in the output.

It is apparent that if the size of the filter capacitor is increased, the average output voltage would rise and the ripple would be reduced. This, however, will increase the diode pulse current and this factor must be considered. If the capacitor size was increased to 100 microfarads (in Example 5.1), the load voltage for the 1 kilohm load would rise to a slightly higher value than the 3.3 kilohm load with the 22 microfarad capacitor. Now, if the capacitor size was increased to 1000 microfarads the average load voltage would be only just less than the no load voltage. Students should confirm these figures for themselves.

This could be summed up by saying that to almost eliminate ripple (say to 0.1 of 1 per cent) a large capacitor must be used. The size of the capacitor also depends on the load current. Many small power supplies for electronic circuits could employ capacitors of up to say 100 000 microfarads.

5.5 Filter circuits—chokes

The action of a *choke* (more accurately termed an *inductor*), is to oppose any change in current flowing through it (Lenz's Law). Basically, if the current tends to fall, the emf of self induction decreases and tends to keep the current at its original level. The opposite takes place with a rise in current: the emf of self induction increases and opposes the change. This action will tend to prevent a ripple voltage reaching its peak value or falling to its minimum value. This can be seen in Figure 5.13 where the effect of both a capacitor and an inductor are illustrated.

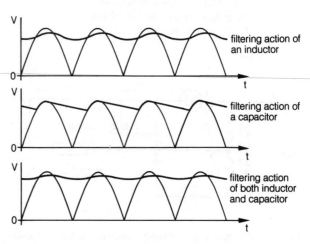

Fig. 5.13 *Filtering action of capacitors and inductors*

Where currents are too high for simple capacitor filters (the capacitor would be far too large) both inductors and capacitors are employed. These filters are often called *choke input filters* and a typical arrangement can be seen in Figure 5.14. We could say that the action of the inductor is to tend to 'level out' the ripple so the capacitor then has an easier task of 'filling in the hollows'. The action of the inductor is that it has little opposition to the flow of steady dc (only its resistance) but has a high impedance to the varying ripple. The capacitor, on the other hand, blocks any dc but is a low impedance path for the varying ripple.

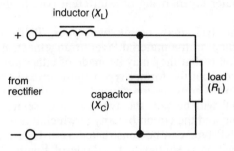

Fig. 5.14 *A choke input filter*

For a given load, when the average load current falls, the effect of the inductor decreases until, at very small currents, it has little effect. The filter then operates as a capacitor filter and the voltage rises up towards the peak value of the wave. This effect increases the regulation of the filter (rise of output voltage with fall in output current).

In the circuit in Figure 5.14, to ensure a minimum acceptable ripple in the output, the following must be observed:

$$X_C \; < \; \frac{R_L}{10} \qquad (5.3)$$

where R_L is the load resistance in ohms,
   X_C is the capacitive reactance of the capacitor in ohms.

Also: $\qquad X_L \; > \; 10\,X_C \qquad (5.4)$

where X_L is the inductive reactance of the inductor in ohms.

When power supplies are very large (for industrial power purposes) inductors alone are used for filtering since, because of the rectifiers used, the ripple is very small to start with.

Unit 5 **SUMMARY**

- A filter is any network which discriminates against frequencies.
- Filters can be low pass, high pass, band pass, or band block.
- Filters used for power supplies are low pass filters.
- Filtering, in power supplies, is the removing of the ripple from the output.
- On small power supplies the most-used filter is the capacitor filter.
- The higher the load current, the larger must be the capacitor for adequate filtering.
- The peak diode current is many times higher than the average load current when a capacitor filter is used.

- At no load, the output voltage equals the V_{AC} peak voltage when a capacitor filter is used.
- In a well-filtered power supply the output voltage is very little less than the V_{AC} voltage, when capacitor filtering is used.
- In medium-sized power supplies a choke input filter may be used.
- A choke, or inductor, opposes any change in current and so tends to 'level out' any ripple in a rectifier output.

Bipolar junction transistors (BJTs)

6.1 Development

The introduction of the bipolar junction transistor (BJT) was one of the most startling developments in the history of electronics—and possibly the one with the most far reaching significance. The first practical transistor was developed in the Bell Telephone Laboratories in 1948; its invention is credited to three physicists: John Bardeen, Walter Brattain, and William Shockley. The development came about from investigations they were making into the physics of the semiconductor diode. The name *transistor* was coined from the two words 'transfer' and 'resistor'.

The term *bipolar* comes from the fact that in BJTs the current is carried by both electrons *and* holes, which are negative and positive current carriers. This addition to the name transistor was necessary because other types of transistors have since been developed which utilise only one type of current carrier.

From the first BJT transistor has developed the huge array of controlled semiconductor devices which may be grouped under the generic title *solid state electronic* devices.

6.2 The three layer device

PN junction diodes are composed of *two* layers, one of P-type material and the other of N-type material, and so may be termed two-layer devices. The BJT, however, is made up of *three* layers of semiconductor material—

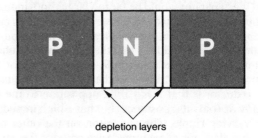

Fig. 6.1 *Representation of a PNP bipolar junction transistor*

both outer layers being of a different type to the centre layer.

The BJT may be of two types, PNP or NPN, depending on the material layer arrangement. A further variation is that they may be made of either germanium or silicon, but the former type is now virtually obsolete. The type of construction depends upon: the use to which the BJT will be put; the type of basic semiconductor element; and the circuit polarity in which it is to be used. The PNP type can be constructed as in Figure 6.1: this construction is similar to the diode of Figure 2.2, with the exception that now there are two junctions, each producing a depletion layer.

Many methods of manufacture will produce the three-layer construction, but whatever method is used the operation of the device is similar. The actual dimensions of the three layers is very much smaller than the representation in Figure 6.1. In practice the centre section is made as thin as possible, and dimensions in the order of 0.05 millimetre are usual. The two outer layers are also kept relatively thin but usually have a larger dimension of thickness than the centre; one of these outer regions is made of a lower resistivity than the other.

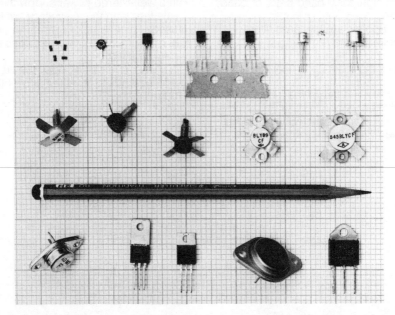

Fig. 6.2 *A range of BJTs for various applications. The upper group are for general purposes, from 0.1 to 1 watt. The central group are for radio frequency power amplification with outputs from 5 to 100 watts. The lower group are power control devices with dissipations up to 150 watts*   PHILIPS ELECTRICAL COMPONENTS AND MATERIALS

6.3 Base, emitter and collector

Whereas in the PN diode the two sections were named anode and cathode, the three sections of the BJT have different names. The thin centre section is referred to as the *base*; the end section with the lower resistivity is termed the *collector*; and the other end is the *emitter*. Leads are attached to each of these sections and the complete unit is encapsulated in one of a number of standard body types, some of which are illustrated in Figure 6.2. It does not matter whether the BJT is an NPN or PNP type: the three sections are given the same name regardless.

The graphic symbols for NPN and PNP BJTs, together with the layer construction alongside, are illustrated in Figure 6.3. Note that the arrow is always the emitter no matter in what direction the arrow is placed. The arrow indicates the direction of conventional current flow (positive to negative) in normal operation. The graphic symbols may be mirror reversed or oriented in any direction without changing their validity.

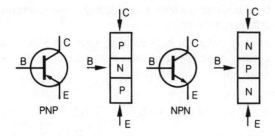

Fig. 6.3 *BJT symbols*

6.4 BJT operation—current paths

The BJT could be said to be two diodes joined together (Fig. 6.4), as there is a PN junction joined to an NP junction. Considering the PNP BJT, the emitter may be considered the anode, and the base the cathode, of the emitter-base junction. Then the base is the cathode, and the collector the anode, of the base-collector junction. (For an NPN BJT the roles of the cathode and anode are reversed.) It must, however, be emphasised that two diodes connected in this way will *not* act as a BJT.

If an NPN BJT is connected to a supply as in Figure 6.5, it will be found that with only the V_{CC} supply connected no current can flow, since the collector-base junction is reverse biased. When the V_{BB} supply is connected as well, by closing switch S, it will forward bias the base-emitter junction, and current will not only flow from the base to the emitter but also from the collector to the emitter. This is known as *transistor action*.

The current distribution in the circuit of Figure 6.5 may be expressed as an equation determined by Kirchhoff's current law:

$$I_E = I_B + I_C \tag{6.1}$$

where I_E is emitter current,
  I_B is base current,
  I_C is collector current.

By measurement it may be determined that the value of I_B is *very* small in relation to I_E and I_C, so an approximation may be made:

$$I_E \simeq I_C \tag{6.2}$$

6.5 Theory of operation

In many courses the theory of operation of the BJT is not discussed as it is really not essential for the understanding of its use. However, for those who feel this is necessary, this section has been included in the book.

Reconsider the circuit of Figure 6.5. Here the collector-base junction is reverse biased, using a 6 volt battery, and the emitter-base junction forward biased, using a 1.5 volt battery. In the emitter-base circuit, a series resistor and rheostat are included to limit and control the current. If switch S is opened, the current in the collector-base circuit (reverse or leakage current) is found to be *very* small, as could be expected (it may be in the order of, say, 5 microamperes). At this point, note that the current, due mostly to thermally released electron-hole pairs, is very dependent on temperature—and should the temperature rise, current will also increase. (As a rough guide, it will double itself for each 10⁰C rise in temperature.)

This current is referred to as I_{CBO}, that is: collector-base current when the emitter-base current is zero. Now

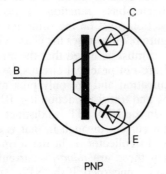

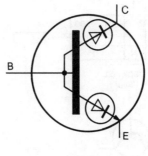

Fig. 6.4 *The two junctions of a BJT may be considered to be two diodes*

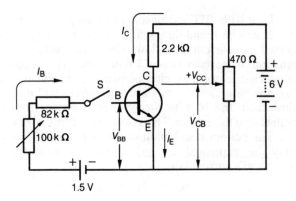

Fig. 6.5 *Current paths in an NPN BJT*

when the switch S is closed and the emitter current I_E adjusted to, say, 1000 microamperes (1 milliampere), the collector current rises to about 980 microamperes. This current consists of the original 5 microamperes leakage plus about 975 microamperes of the 1000 microamperes flowing out from the emitter. The remaining current of 20 microamperes flows into the base (I_B). From this it is seen that $I_E = I_C + I_B$. This is very important and remains true no matter how the BJT is used.

If the collector-base voltage (V_{CB}) is varied, there is negligible change in I_C. This is because of the very high apparent resistance of the collector circuit in the reverse biased condition.

With reference to Figure 6.6, it is apparent that in the normal state there is a depletion layer at each PN junction. These are marked 1 and 2. As the collector-base junction is reverse biased the depletion layer is widened in a similar manner to the PN diode in Figure 2.5. Now the base (in an actual BJT) is so thin that this depletion layer extends well into it, and so the depletion layer's effect is very close to the emitter end.

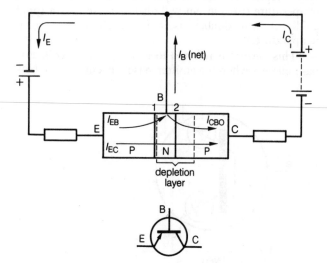

Fig. 6.6 *Internal and external current paths in a BJT with the BJT symbol below*

Since the emitter-base junction 1 is forward biased (as in Fig. 6.6) forward current flows. This current is in the form of holes which are injected into the base region from the P-type emitter. The important point to note is that as the base is so thin and the depletion layer of junction 2 so wide, holes are swept right across junction 2 by the negative polarity of the collector. A current of holes then exists between emitter and collector. V_{EB} controls the magnitude of the hole current and V_{CB} takes up the current as soon as the emitter supplies it.

There is some recombination of holes in the base region with electrons flowing up from the base lead. Also, electrons flow from the emitter around the circuit to the collector to maintain the hole current through the junction.

There are three current effects in the transistor: the emitter-collector current within the device; the emitter-base current; and I_{CBO} between base and collector.

Now I_{CBO} consists of holes in the collector P-type material and electrons in the N-type base. These electrons in the base region combine with some of the holes from the emitter (i.e. recombination) and so actually reduce the base-emitter current. Considering these currents as percentages, we have:

emitter-collector hole current, approx. 98 per cent;
emitter-base hole current, approx. 2 per cent;
I_{CBO}, approx. 0.5 per cent.

Now:

net base lead current (I_B)	=	2–0.5
	=	1.5 per cent
net collector current (I_C)	=	98+0.5
	=	98.5 per cent
net emitter current (I_E)	=	98.5+1.5
	=	100 per cent

Note that the main current through the base region is by holes which are *minority* current carriers as far as the N-type base is concerned.

Although the reference so far has been to PNP BJTs, the current flow in the NPN BJT is the same. The only difference is that the current carriers and direction of flow in the NPN BJT are exactly the opposite of those in the PNP type. The most important point is that the polarities of the batteries *must* be reversed so that the emitter-base junction is still forward biased and the collector-base junction is reverse biased. The bias direction must be the same for both types of BJTs and Figure 6.7 illustrates the change in polarity for each type.

It must be noted that the circuits depicted in Figure 6.7 are not practical circuits, i.e. they have no practical application and are only presented to illustrate current flow and transistor action in a BJT.

In the NPN BJT (as distinct from the PNP type) the main current through it, that is from emitter across the base to collector, is in the form of electrons, and once again these are *minority* current carriers as far as the base is concerned. If the NPN BJT is replaced by a PNP type in Figure 6.6, the batteries would have to be reversed

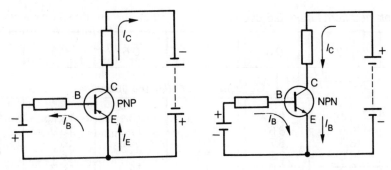

Fig. 6.7   *Polarity connection for PNP and NPN BJTs*

in polarity, and the current direction would change. Also, 'holes' would become 'electrons' and 'electrons' would become 'holes'.

One very important point to remember is that the arrowheads in the graphical symbols represent the direction of *conventional* current flow, and *electron* flow is in the opposite direction. *Hole* flow is in the same direction as conventional current flow.

In many circuits it is necessary to use both PNP *and* NPN BJTs. One such application is the output power BJTs in some amplifiers, where the BJTs are connected in what is termed *complementary symmetry*. This means the two BJTs are identical in their electrical characteristics except one is PNP and the other NPN. Students, at this stage, should not be concerned with these terms except to know that such special complementary BJTs are manufactured.

6.6 Current control in the BJT

It has been established that a current must flow in the base-emitter circuit before any appreciable current can flow in the collector-emitter circuit. There is a definite relationship between the magnitude of base current (I_B)

and collector current (I_C) in a given BJT, over a given range of current. It is called the *current gain* of the BJT. It means that, if a small current flows in the base circuit, it will *permit* a larger current to flow in the collector circuit.

The ability of the base current to control the collector current, or the current gain of the BJT, is represented by the symbol h_{FE} (the symbol β, the Greek letter beta, is also used in which case, strictly, it should be β_{DC}). It can then be said that the relationship between the *unvarying* base current and the collector current may be expressed as:

$$I_C = h_{FE} I_B \qquad (6.3)$$

where I_C is collector current,
  I_B is base current, and
  h_{FE} is current gain of the BJT.

It may also be said that the relationship of collector current to base current is the current gain of the BJT, and Equation 6.3 may be transposed to:

$$h_{FE} = \frac{I_C}{I_B}$$

This is often referred to as the dc current gain of the BJT (also expressed as β_{DC}).

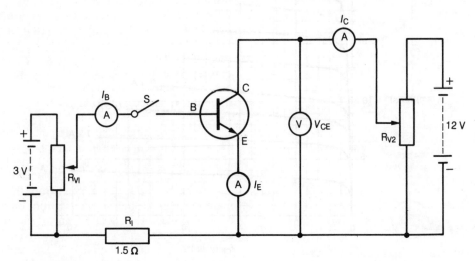

Fig. 6.8   *BJT investigation circuit*

Table 6.1 *Test results obtained from Figure 6.8 circuit*

I_B (amperes)	0.1	0.2	0.3	0.4	0.6	0.8	1.0	1.2
V_{CE} (volts)	Collector current (I_C) in amperes							
0	0	0	0	0	0	0	0	0
1	3.49	6.45	8.5	10.25	13.1	15.75	17.75	19.5
2	3.5	6.5	8.6	10.5	13.3	15.9	17.9	19.75
3	3.52	6.55	8.7	10.6	14.0	16.1	18.1	20.0
4	3.55	6.6	8.9	10.8	14.1	16.2	18.25	
5	3.6	6.65	9.0	11.0	14.3	16.3		
6	3.75	6.7	9.15	11.25	14.5			
7	3.8	6.75	9.3	11.5				
8	3.9	6.8	9.45					
9	3.95	6.9						
10	4.0							

6.7 Characteristic curves

If a BJT is set up in a test circuit as shown in Figure 6.8 (p. 43), both the base-emitter voltage and collector emitter voltage supplies may be varied. With these variations both the base current and collector current may be monitored, the figures giving us the current gain of the BJT. But more information than that can be obtained.

The base current can be set to a certain value and the collector-emitter voltage (V_{CE}) varied from zero to maximum. At each variation the collector current (I_C) can be noted. If this procedure is repeated for each setting of the base current we obtain the results in Table 6.1.

From these readings a family of curves may be drawn, as shown in Figure 6.9. Note that each curve is plotted from the same information, that is variation of I_C with change in V_{CE}. Each curve, however, is drawn for a particular I_B, ranging here from $I_B = 0.1$ A, to $I_B = 1.2$ A.

An examination of these curves will show that below about 0.8 volt there is a very rapid fall of collector current, with a reduction in collector-emitter voltage. This is known as the saturation region, in which the collector-base junction is effectively forward biased, and transistor action virtually ceases. This region of the curve is of importance when the BJT is used as a switch but is useless when the effect of current gain is required.

It can also be seen from the figure that the curves are virtually linear from the 'knee' and that there is very little change in collector current with a change in collector-emitter voltage, due to the very high resistance of the reverse biased collector-emitter junction.

These curves were produced from a test on a power BJT. BJTs can be small signal types with a maximum collector current of about 20 milliamperes, or power types with collector currents of many amperes.

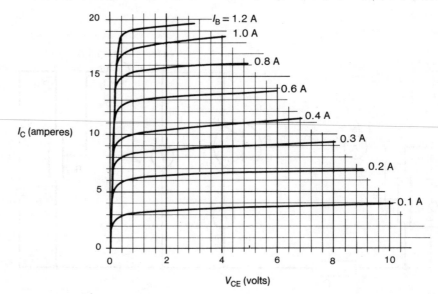

Fig. 6.9 *BJT collector characteristic: family of curves*

The reading of collector currents at the higher base current values are not taken at the higher voltage values, because the power dissipation in the BJT must be limited to a maximum value to keep the junction temperature below a maximum value. When operated with a *heat sink*, a junction temperature of no greater than 90⁰C is allowed for this particular type of BJT. If this temperature is exceeded, the BJT may be destroyed.

6.8 The BJT as a switch—cut-off

In Section 6.4 it was mentioned that with zero base current, as in Figure 6.5(a), no collector current will flow in the BJT circuit. This condition is known as *cut-off*, and referring to Figure 6.10 we will now examine the voltages and current about the circuit.

If V_1 is zero and as a consequence I_B is zero, I_C will also be zero by transistor action (for all practical purposes)—so there will be no voltage drop across R_C, that is:

$$I_C R_C = 0$$

Then, by Kirchhoff's Voltage Law, the full supply voltage V_{CC} will appear across the BJT or, expressed as an equation:

$$V_{CE} = V_{CC}$$

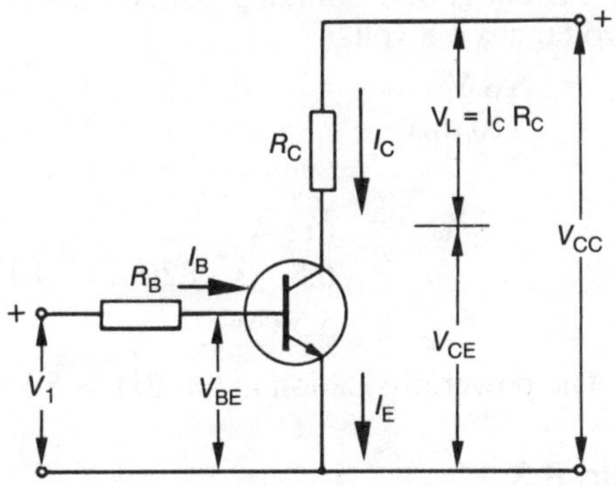

Fig. 6.10 *Current and voltage relationships in a BJT switch circuit*

6.9 The BJT as a switch—saturation

The BJT is said to be *saturated* when the voltage drop across the BJT between collector and emitter is zero (or close to zero, in a practical circuit), that is: $V_{CE} \rightarrow 0$. In this case the full supply voltage will appear across R_C (in Fig. 6.10) and by Kirchhoff's Voltage Law may be expressed as:

$$I_C R_C = V_{CC} - V_{CE}$$
$$\text{and so } I_C R_C \simeq V_{CC}$$

There is naturally a certain maximum value of I_C that will produce saturation, and this is obtained from the Ohms's Law equation:

$$I_{C(MAX)} = \frac{V_{CC}}{R_C} \qquad (6.4)$$

where $I_{C(MAX)}$ is maximum collector current,
   V_{CC} is supply voltage, and
   R_C is collector load resistance.

To find the minimum base current to produce saturation, we apply the current gain equation, that is:

$$I_{B(MIN)} = \frac{I_{C(MAX)}}{h_{FE}} \qquad (6.5)$$

and, substituting equation 6.4 in equation 6.5:

$$I_{B(MIN)} = \frac{V_{CC}}{h_{FE} R_C} \qquad (6.6)$$

This is termed the *minimum* value of I_B to produce saturation, since in values of base current above this, the relationship $\frac{I_C}{I_B} = h_{FE}$ does not hold because I_C is unable to increase. That is, the BJT is saturated and cannot pass any more collector current in the given circuit.

To determine the voltage V_1 to produce saturation, it can be shown that the following equation is sufficiently accurate:

$$V_{1(MIN)} = \frac{V_{CC} R_B}{h_{FE} R_C} \qquad (6.7)$$

where $V_{1(MIN)}$ is minimum input voltage to produce saturation,
   V_{CC} is supply voltage,
   R_B is input base resistance,
   h_{FE} is current gain,
   R_C is called collector load resistance.

The value of h_{FE} used must be the minimum given by the BJT manufacturer for the BJT in use.

Example 6.1

If in the circuit of Figure 6.10 the supply voltage (V_{CC}) is 18 volts, the minimum h_{FE} of the BJT is quoted as 80, the collector load resistor (R_C) has a resistance of 1000 ohms and the input base resistor (R_B) has a resistance of 22 kilohms, determine:

(a) maximum value of I_C;
(b) minimum value of I_B for saturation;
(c) minimum value of V_1 for saturation.

$$V_{CC} = 18 \text{ V}$$
$$h_{FE(MIN)} = 80$$
$$R_C = 1 \text{ k}\Omega$$
$$R_B = 22 \text{ k}\Omega$$

(a) $$I_{C(MAX)} = \frac{V_{CC}}{R_C}$$
$$= \frac{18}{10^3}$$
$$= 18 \times 10^{-3}$$
$$= 0.018 \text{ A}$$

Answer (a): The maximum value of I_C is 18 milliamperes.

(b)
$$I_B = \frac{I_C}{h_{FE}}$$
$$= \frac{18 \times 10^{-3}}{80}$$
$$= 2.25 \times 10^{-4}$$
$$= 0.000225 \text{ A}$$

Answer (b): The minimum value of base current for saturation is 0.225 milliamperes.

(c)
$$V_1 = \frac{V_{CC} R_B}{h_{FE} R_C}$$
$$= \frac{18 \times 22 \times 10^3}{80 \times 10^3}$$
$$= 4.95 \text{ V}$$

Answer (c): The minimum input voltage for saturation is 4.95 volts.

6.10 BJT switch circuits

Considering the circuit of Figure 6.11 we have a pilot lamp being switched by a BJT (actually this is a composite of two BJTs, called a *Darlington*, but this need not concern us here). The input voltage is 2.5 volts but the input current is only about 0.35 milliampere, which represents an input power of only about 0.9 milliwatt. As the power controlled is in the order of 3 watts we could say that the power gain of the circuit is about 3000 or, in other words, the input is able to control a power of 3000 times its own value.

The Darlington BJT in Figure 6.11 has a current gain of about 750 (h_{FE}). From the information in Section 6.8 confirm the approximate conditions for the circuit.

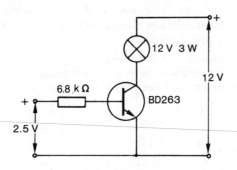

Fig. 6.11 *A practical BJT switch circuit*

6.11 Power ratings of BJTs

As has been mentioned in previous units, BJTs are used for many different purposes and many hundreds of different types are manufactured. Broadly speaking they can be divided into four main types:

1. small signal low frequency;
2. power low frequency;
3. high frequency;
4. switching.

As this book does not cover high frequency applications (radio, TV and the like) and switching BJTs are mainly limited to computer and similar applications, the two types which interest us are small signal and power types.

There is really no hard and fast rule of what constitutes a small signal and power BJT, but a general rule is that BJTs with ratings of over about 1 watt may be classified as power types.

The power rating of a BJT is the maximum power that may be absorbed within the semiconductor material and at the junctions without the junction temperature rising above a certain maximum value. The power dissipated within the BJT can be calculated from the collector current and collector-emitter voltage:

$$P_{tot} = V_{CE} I_C \qquad \textbf{(6.8)}$$

where P_{tot} is the power dissipation in the BJT in watts,
V_{CE} is the collector-emitter voltage in volts,
I_C is the collector current in amperes.

Example 6.2
Determine the power dissipated in a BJT when the collector current is 820 milliamperes and the collector-emitter voltage is 6.8 volts.
$$V_{CE} = 6.8 \text{ V}$$
$$I_C = 820 \text{ mA}$$
$$P_{tot} = ?$$
$$P_{tot} = V_{CE} I_C$$
$$= 6.8 \times 820 \times 10^{-3}$$
$$= 5.58 \text{ W}$$

Answer: The power dissipation in the BJT is 5.58 watts.

Example 6.3
Determine the power dissipation in a BJT when the collector current is 1.5 milliamperes and the collector-emitter voltage is 6.6 volts.
$$V_{CE} = 6.6 \text{ V}$$
$$I_C = 1.5 \text{ mA}$$
$$P_{tot} = ?$$
$$P_{tot} = V_{CE} I_C$$
$$= 6.6 \times 1.5 \times 10^{-3}$$
$$= 0.0099 \text{ W}$$

Answer: The power dissipation in the BJT is 0.01 watt or 10 milliwatts.

From these two examples it can be seen that the first BJT is a power type and the second a small signal type.

Power BJTs are made in ratings up to hundreds of watts but generally a rating of about 100 watts is the maximum in common use.

6.12 List of terms connected with BJTs

I_B **base current** Current entering or leaving the base terminal.

I_C **collector current** Current entering or leaving the collector terminal.

I_E **emitter current** Current entering or leaving the emitter terminal.

I_{CB0} **collector-base leakage current** Minority carrier current that flows across the reverse biased collector base junction.

I_{CE0} **minimum collector-emitter current** The current that flows between collector and emitter when the base current is zero.

h_{fe} **or** β The ac current gain.

h_{FE} **or** β_{DC} The dc current gain.

P_{tot} **maximum power rating** The maximum power that can be absorbed by the BJT without the junction temperature rising above a maximum safe value.

V_{CE} **collector-emitter voltage** Voltage between the collector and emitter terminals.

V_{CEO} **maximum collector emitter voltage** The maximum voltage between the collector and emitter terminals with the base open-circuited. If this voltage is exceeded the BJT may be destroyed.

V_{EBO} **maximum reverse emitter-base voltage** The maximum reverse voltage that can be applied between emitter and base terminals without damaging the BJT.

Unit 6 **SUMMARY** (Sections 6.1 to 6.10 only)

- In a bipolar junction transistor (BJT) current carriers are *both* holes and electrons.
- A BJT is made from three layers of doped semiconductor, either PNP or NPN.
- The three terminals of a BJT are termed collector, base and emitter.
- In a BJT the emitter current is always equal to the base current plus the collector current, but as the base current can be very small we can say that, approximately, emitter current equals collector current.
- The flow of collector current as a result of the flow of base current is called transistor action.

- The ratio of collector current to base current is called the current gain of a BJT. The symbol for current gain is h_{FE} or β (beta).
- When no base current flows a BJT is said to be cut off. In this condition no collector current flows.
- When a BJT is saturated it can pass no more collector current and the voltage drop across it (V_{CE}) is very low.
- The power lost as heat in a BJT is a product of collector current and collector-emitter voltage. At cut off no power is lost and at saturation very little power is lost.

The following sections in this unit are only for those students who wish to study above the normal level of this book.

Current gain in the BJT

6.13 DC current gain

In Section 6.6, the current control of the BJT, which is the ability of the current in the base circuit to control the magnitude of current in the collector circuit, was examined. It was found that the relationship between base current and collector current was termed *current gain*. This current gain was expressed as the ratio of collector current to base current and expressed mathematically as:

$$h_{FE} = \frac{I_C}{I_B}$$

where h_{FE} is termed the dc current gain of the BJT.

The term *dc gain* is used because we are considering a certain fixed, or unvarying, base current which will permit a fixed unvarying collector current to flow. This is the fundamental operation of the BJT.

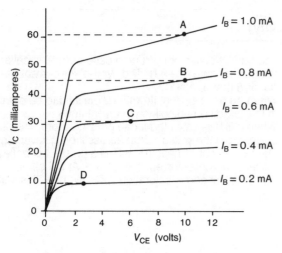

Fig. 6.12 *Collector output characteristics of a BJT*

The dc current gain (h_{FE}) is not a constant value over the operating range of a BJT. At low collector currents it may be smaller than at high collector currents. For this reason most manufacturers state the dc current gain at a specific collector current. In addition, the collector current does increase slightly with an increase in collector-emitter voltage (i.e. the collector characteristic curves have a small positive slope) and this is greater at higher collector currents and, by inference, higher base currents.

In Figure 6.12 four points are marked on the family of curves: A,B,C,D. The dc current gain of the BJT will be determined for each point.

At point A, with a base current of 1 milliampere, and at a collector-emitter voltage of 10 volts, the collector current is 60 milliamperes. Using equation 6.3:

$$
\begin{aligned}
I_B &= 1 \text{ mA} \\
I_C &= 60 \text{ mA} \\
h_{FE} &= ?
\end{aligned}
$$

$$
\begin{aligned}
h_{FE} &= \frac{I_C}{I_B} \\
&= \frac{60}{1} \\
&= 60
\end{aligned}
$$

Answer: The dc gain is 60.

At point B:

$$
\begin{aligned}
I_B &= 0.8 \text{ mA} \\
I_C &= 45 \text{ mA} \\
h_{FE} &= ?
\end{aligned}
$$

$$
\begin{aligned}
h_{FE} &= \frac{I_C}{I_B} \\
&= \frac{45}{0.8} \\
&= 56
\end{aligned}
$$

Answer: The dc gain is 56.

At point C:

$$
\begin{aligned}
I_B &= 0.6 \text{ mA} \\
I_C &= 30 \text{ mA} \\
h_{FE} &= ?
\end{aligned}
$$

$$
\begin{aligned}
h_{FE} &= \frac{I_C}{I_B} \\
&= \frac{30}{0.6} \\
&= 50
\end{aligned}
$$

Answer: The dc gain is 50.

At point D:

$$
\begin{aligned}
I_B &= 0.2 \text{ mA} \\
I_C &= 9 \text{ mA} \\
h_{FE} &= ?
\end{aligned}
$$

$$
\begin{aligned}
h_{FE} &= \frac{I_C}{I_B} \\
&= \frac{9}{0.2} \\
&= 45
\end{aligned}
$$

Answer: The dc gain is 45.

A manufacturer would probably quote the gain at a V_{CE} of about 6 volts and a collector current of 30 milliamperes. This would be about point C on the curves of Figure 6.12 and so the quoted gain would be 50.

It must also be pointed out that in mass production it is virtually impossible for a manufacturer to construct every BJT with the same gain. In practice, there is a spread of gains and it is quite possible to have the gain of a single type number BJT varying from 100 to 800 for individual BJTs. Also quite often BJTs of the same type number are separated into sub-groups. As an example, one manufacturer could group type BC108 into sub-groups A, B and C. The complete type number with their range of dc current gains could be: BC108A, 110–220; BC108B, 200–450; BC108C, 420–800.

6.14 AC current gain

When the *change* in base current produces a correspondingly larger *change* in collector current, the relationship between these two *changes* is termed the *ac current gain*. Mathematically it is the ratio of change in collector current to change in base current, collector-emitter voltage being constant. This may be expressed as:

$$h_{fe} = \frac{\Delta I_C}{\Delta I_B} \bigg| \ V_{CE} \text{ being constant} \qquad \textbf{(6.9)}$$

where h_{fe} is the ac current gain,
ΔI_C is the change in collector current,
ΔI_B is the change in base current, and
V_{CE} is collector-emitter voltage.

The ac gain of a BJT may be determined from the collector characteristic curves or from a current gain curve. The collector characteristic curves (similar to the curves in Section 6.7) show the variation in collector current to collector-emitter voltage at differing values of base current. To determine the ac current gain from these curves, it is first necessary to determine a fixed value of V_{CE}. In Figure 6.13 this has been fixed at 9 volts. A line is drawn vertically from the 9 volt V_{CE} point to cut three curves about point A, which is arbitrarily chosen about the centre of the curves. A line is now drawn from where the 9 volt V_{CE} line cuts the two curves, $I_B = 0.4$ mA and $I_B = 0.8$ mA, to the I_C axis. Now the change in I_C is the scaled distance between these lines

on the I_C axis and the change in I_B is the difference between the two I_B curves.

In Figure 6.13 the change in I_C is about 26 milliamperes and the change in I_B is simply the difference between the $I_B = 0.8$ mA and the $I_B = 0.4$ mA curves, 0.4 milliamperes. Placing these two figures in equation 6.9 we get:

$$\Delta I_C = 26 \text{ mA}$$
$$\Delta I_B = 0.4 \text{ mA}$$
$$h_{fe} = ?$$

$$h_{fe} = \frac{\Delta I_C}{\Delta I_B} \bigg| \ V_{CE} \text{ constant}$$
$$= \frac{26}{0.4}$$
$$= 60$$

Answer: The ac gain is 60.

The current gain curve may be constructed from the collector characteristic curves. A value of V_{CE} is chosen, in Figure 6.13 this is 9 volts, and the collector currents where this V_{CE} line cuts the base current curves are plotted on a graph of collector current versus base current drawn in the second quadrant, as in Figure 6.13. The curve obtained from these plots is the gain curve (for the particular V_{CE} chosen).

When the gain curve is drawn, a triangle is constructed on the curve about the operating point B, which is the same value as point A, so that two sides form a right angle with the curve as the hypotenuse. The vertical side represents the change in I_C and the horizontal side the change in I_B. These two values are then scaled off their respective axes. From the curve in Figure 6.13 ΔI_C is 28.2 milliamperes and ΔI_B is 0.47 milliampere. Again using equation 6.9 we get:

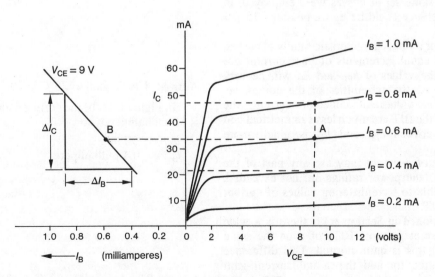

Fig. 6.13 *Current gain curve (left) and collector characteristics curves (right) of a BJT*

$$\Delta I_C = 28.2 \text{ mA}$$
$$\Delta I_B = 0.47 \text{ mA}$$
$$h_{fe} = ?$$

$$h_{fe} = \left.\frac{\Delta I_C}{\Delta I_B}\right| V_{CE} \text{ constant}$$

$$= \frac{28.2}{0.47}$$

$$= 60$$

Answer: The ac gain is 60.

Another term for ac current gain (h_{fe}) is β (the Greek letter beta) and the student should be aware of the alternative symbol.

Example 6.4

In a certain BJT a change in base current of from 2 to 2.6 milliamperes produces a change in collector current of from 260 to 326 milliamperes, when V_{CE} is held constant. What is the ac current gain of the BJT?

$$I_C = 260 - 326 = 66 \text{ mA}$$
$$I_B = 2.6 - 2 = 0.6 \text{ mA}$$
$$h_{fe} = ?$$

$$h_{fe} = \left.\frac{I_C}{I_B}\right| V_{CE} \text{ constant}$$

$$= \frac{66}{0.6}$$

$$= 110$$

Answer: The ac gain is 110.

6.15 Comparison between dc and ac gain

The ac and dc gains of a single type BJT (Sections 6.11 and 6.12) do come out to a different value. This is despite the fact that the same set of curves were employed. In practice this difference could be up to, possibly, 15 per cent.

In the collector output characteristic family of curves, if the curves for equal increments of base current are equally spaced, the values of h_{FE} and h_{fe} will be very similar in value. At the extremities of the curves, i.e. at high and low values of collector current, the characteristics of the BJT are much less symmetrical and the difference between h_{FE} and h_{fe} is much more pronounced.

The designer of a circuit may use any part of the curves, as long as the power ratings are not exceeded, so that it is possible to have differing values of current gain in a given BJT.

As was mentioned in Section 6.11, there is a wide spread in the current gains of BJTs of even the same type number and this is quite normal. This difference *can* be compensated for and the actual current gain becomes much less important.

Although there may be a difference between h_{FE} and h_{fe}, in practice this difference can really be ignored and the centre dc current gain employed. In fact, most manufacturers use only the dc current gain in their published data.

As mentioned in Section 6.12 the alternative term for h_{fe} is β; also the symbol h_{FE} is sometimes replaced with β_{DC}. However for most practical purposes we can write the current gain equation simply as:

$$\beta = \frac{I_C}{I_B}$$

and this is sufficiently accurate for most work.

6.16 Effects of temperature on current gain

It was seen that in the case of diodes the reverse leakage current increased with an increase in temperature. In the case of BJTs, the internal collector-base current (I_{CBO}) will also increase with an increase in temperature. This of course is an increase in minority current carriers and so allows the collector to capture more current carriers for a given base current. It can be seen then that an increase in temperature will produce an increase in current gain.

Figure 6.14 illustrates the difference in collector characteristics of the same BJT at different temperatures. The effect of the increase in temperature is virtually to lift the curves up the vertical I_C axis.

To illustrate the change in h_{FE} with change in temperature, consider a base current of 0.4 milliampere at a V_{CE} of 6 volts in both sets of curves.

In Figure 6.14(a), the corresponding collector current is 20 milliamperes, so:

$$I_C = 20 \text{ mA}$$
$$I_B = 0.4 \text{ milliamperes}$$
$$h_{FE} = ?$$

$$h_{FE} = \frac{I_C}{I_B}$$

$$= \frac{20}{0.4}$$

$$= 50$$

Answer: The dc gain at 25°C is 50.

In Figure 6.14(b) the corresponding collector current is 50 milliamperes, so:

$$I_C = 50 \text{ mA}$$
$$I_B = 0.4 \text{ milliampere}$$
$$h_{FE} = ?$$

$$h_{FE} = \frac{I_C}{I_B}$$

$$= \frac{50}{0.4}$$

$$= 125$$

Answer: The dc gain at 100°C is 125.

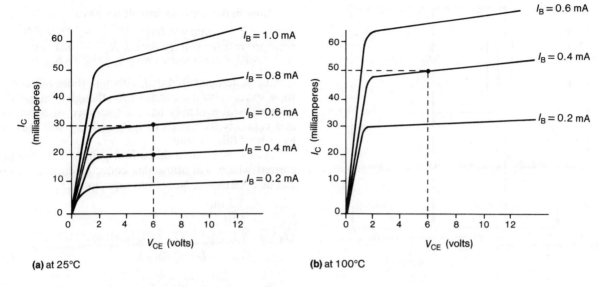

Fig. 6.14 *BJT collector characteristics at different temperatures*

From the two foregoing examples it can be seen that the value of h_{FE} (or β) has increased and so a certain fixed value of base current will now produce a greater value of collector current. In other words it may be said that an increase in temperature will produce a rather significant increase in collector current, with a constant base current. This effect can be quite serious under certain conditions.

6.17 Input characteristic

The base-emitter junction of a BJT can be considered to be a forward biased PN junction, similar to a PN diode. An increase in base-emitter voltage past the 'knee' of the curve will produce a linear relationship between base current and base-emitter voltage. As has been stated previously, the linear relationship (the overcoming of the inbuilt depletion layer potential) is about 0.6 volt. This curve is known as the input characteristic and is illustrated in Figure 6.15.

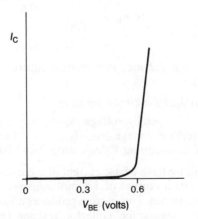

Fig. 6.15 *Input characteristic for a silicon BJT*

As with the output characteristics, the input characteristic will change with variation in temperature. As this variation is in the order of about 2.5 millivolt/°C it can be quite significant. In Figure 6.16 this change in base-emitter voltage for a given base current can be seen.

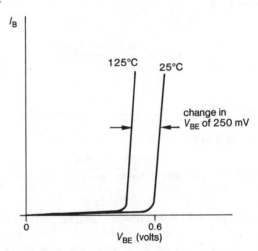

Fig. 6.16 *Change of input characteristic with increase in temperature for a silicon BJT*

6.18 Analysis of a simple circuit

The operation of a BJT in a simple circuit will now be examined. Reference to Figure 6.17 will show a simple biased circuit, similar to that in Figure 6.10. In this circuit the bias current is provided by resistor R_B and the collector current flows through resistor R_C. In this circuit the collector current will be assumed to be 1.5 milliamperes, the supply voltage 15 volts and the BJT to be a silicon type with an h_{FE} of 220. It is required that V_{CE} be about 6.8 volts.

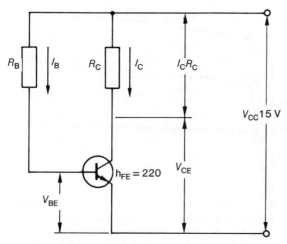

Fig. 6.17 *A simple biased BJT circuit*

Let us first consider the two voltage loops which, by Kirchhoff's Law, are:

$$V_{CC} = I_C R_C + V_{CE} \qquad (6.10)$$
$$V_{CC} = I_B R_B + V_{BE} \qquad (6.11)$$

First let us calculate the value of R_C. Now from equation 10.3

$$I_C R_C = V_{CC} - V_{CE}$$

and

$$R_C = \frac{V_{CC} - V_{CE}}{I_C}$$

now,

$$V_{CC} = 15 \text{ V}$$
$$V_{CE} = 6.8 \text{ V}$$
$$I_C = 1.5 \text{ mA}$$

$$\text{so} \quad R_C = \frac{15 - 6.8}{1.5 \times 10^{-3}}$$
$$= 5467 \ \Omega$$

Answer: Use a 5.6 kilohm resistor (nearest preferred E12 value).

As the value of the resistor chosen is somewhat higher than the calculated value (through necessity) and if we desire the same collector current, the value of V_{CE} may be slightly altered. We will now examine this calculation. By transposing:

$$V_{CC} = 15 \text{ V}$$
$$I_C = 1.5 \text{ mA}$$
$$R_C = 5.6 \text{ k}\Omega$$
$$V_{CE} = ?$$

$$V_{CE} = V_{CC} - I_C R_C$$
$$= 15 - (1.5 \times 10^{-3} \times 5.6 \times 10^3)$$
$$= 6.6 \text{ V}$$

Answer: The collector-emitter voltage is 6.6 volts.

Now in the collector circuit we have:

supply voltage (V_{CC}) = 15.0 volts,
collector resistor voltage drop $I_C R_C$ = 8.4 volts,
collector-emitter voltage V_{CE} = 6.6 volts.

It must be remembered that, although these calculations show exact values, the resistor R_C, though nominally 5.6 kilohms, will probably have a tolerance of 5 per cent and naturally the values of $I_C R_C$ and V_{CE} will also vary by at least this amount.

As the BJT has a dc current gain of 220, the base current which will allow this collector current to flow can be determined. Now from Equation 6.3

$$I_C = 1.5 \text{ mA}$$
$$h_{FE} = 220$$
$$I_B = ?$$

$$I_C = h_{FE} I_B$$

$$\text{so} \quad I_B = \frac{I_C}{h_{FE}}$$

$$= \frac{1.5 \times 10^{-3}}{220}$$

$$= 6.8 \times 10^{-6} \text{A}$$

Answer: The base current must be 6.8 microamperes.

Now, knowing the base current we can calculate the value of the base resistor (R_B) which will permit this current to flow.

From equation 6.11

$$V_{CC} = I_B R_B + V_{BE}$$

and

$$I_B R_B = V_{CC} - V_{BE}$$

so

$$R_B = \frac{V_{CC} - V_{BE}}{I_B}$$

now,

$$V_{CC} = 15 \text{ V}$$
$$I_B = 6.8 \ \mu\text{A}$$
$$V_{BE} = 0.6 \text{ V}$$

(because the BJT is a silicon type)

$$R_B = ?$$

$$\text{so } R_B = \frac{15 - 0.6}{6.8 \times 10^{-6}}$$

$$= 2\,120\,000 \ \Omega$$

Answer: Use a 2.2 megohm resistor (nearest preferred E12 value).

Now in the base circuit we have,

supply voltage (V_{CC}) = 15.0 volts,
base resistor voltage drop ($I_B R_B$) = 14.4 volts,
silicon BJT base-emitter voltage drop = 0.6 volt.

Overall, we have a base current of 6.8 microamperes which permits a current of 1.5 milliamperes to flow in the collector circuit. This current produces a voltage drop of 8.4 volts across the collector resistor (R_B) which produces a voltage of 6.6 volts across the BJT.

6.19 Power ratings of BJTs

When the BJT is operating at near saturation, and V_{CE} is relatively low, a much higher current can pass through it. However, when V_{CE} rises, the current must be a much lower value to keep the power dissipation below the maximum values. In the curves of Figure 6.18, which are the collector characteristic curves for a high power BJT (type 2N3055), a power curve has been drawn over the family of curves. This is a power dissipation curve of 80 watts. The manufacturers state that the 2N3055 BJT has a maximum power rating of 115 watts at a case temperature of 25⁰C. As these are absolute maximum values, the BJT has been derated for safety reasons.

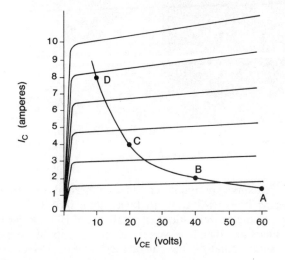

Fig. 6.18   *BJT maximum power dissipation curve*

The points (A,B,C,D) on the curve in Figure 6.18 were determined from the power dissipation equation and the curve drawn in to fit the plotted points. The calculations are as follows:

When V_{CE} = 60 volts, point A,

$$I_C = \frac{P_D}{V_{CE}}$$

$$= \frac{80}{60}$$

$$= 1.3$$

Answer: Maximum collector current at a V_{CE} of 60 volts is 1.3 amperes.

When V_{CE} = 40 volts, point B,

$$I_C = \frac{80}{40}$$

$$= 2$$

Answer: Maximum collector current at a V_{CE} of 40 volts is 2 amperes.

When V_{CE} = 20 volts, point C,

$$I_C = \frac{80}{20}$$

$$= 4$$

Answer: Maximum collector current at a V_{CE} of 20 volts is 4 amperes.

When V_{CE} = 10 volts, point D,

$$I_C = \frac{80}{10}$$

$$= 8$$

Answer: Maximum collector current at a V_{CE} of 20 volts is 8 amperes.

It is obvious that, for the power dissipation to be kept below maximum desired values, the BJT must be operated within the ratings to the left and below the maximum power dissipation curve in Figure 6.18.

6.20 Heatsinks for BJTs

It is essential that the temperature of BJTs be kept below a certain maximum value otherwise they may be destroyed. This is achieved in power BJTs by mounting the BJT on a *heatsink*. (This is also true for power diodes and thyristors.) A heatsink is usually made of extruded aluminium with fins and painted black for maximum radiation. The heatsink dissipates to the surrounding air the heat generated within the device mounted on it. (See Fig. 6.19.)

Manufacturers rate heatsinks in *thermal resistance* which is expressed as degrees per watt (⁰C/W). Thermal resistance is the rate at which a heatsink can dissipate heat to the surrounding air for a given power input for a given temperature rise. To determine the correct thermal resistance heatsink to use, a simplified equation may be employed.

$$R_{TH} = \frac{(T_{j\,MAX} - T_{AMB\,MAX})}{P_D} \qquad (6.12)$$

where R_{TH} is thermal resistance from heatsink to ambient in ⁰C W⁻¹ (*note:* ambient is the surrounding air),

$T_{j\,MAX}$ is the maximum allowed junction temperature of the BJT in ⁰C,

$T_{AMB\,MAX}$ is the maximum temperature reached by the ambient in ⁰C,

P_D is the power dissipation in the BJT in watts.

Example 6.5

Determine the thermal resistance of a heatsink on which must be mounted a BJT which will dissipate 80 watts maximum. The maximum allowed junction temperature of the BJT is 90⁰C and the maximum ambient temperature will be 40⁰C.

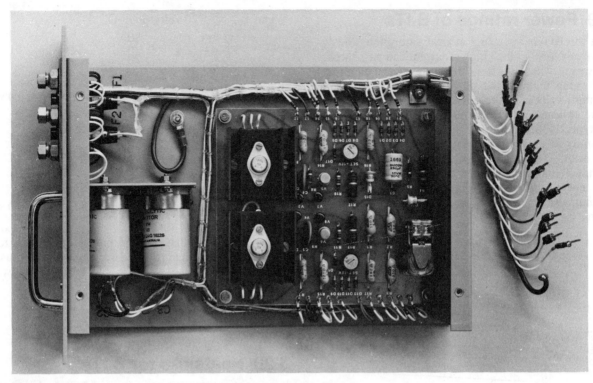

Fig. 6.19 *Light dimming module using BJTs. Note the heatsinks on the power BJTs*

$$T_{j\ MAX} = 90^0C$$
$$T_{AMB\ MAX} = 40^0C$$
$$P_D = 80\ W$$
$$R_{TH} = ?$$

$$R_{TH} = \frac{(T_{j\ MAX} - T_{AMB\ MAX})}{P_D}$$
$$= \frac{90 - 40}{80}$$
$$= 0.625$$

Answer: Use a heatsink with a thermal resistance not exceeding 0.625^{0}C/W.

Heatsinks are manufactured with thermal resistances from as high as 2^{0}C/W to lower than 0.05^{0}C/W. To increase the power ratings of small signal BJTs, a 'push-on' heatsink, which is simply a spring clip of copper with small radiating fins, can be used. This could increase the normal power dissipation of say 10 milliwatts to 50 milliwatts.

Unit 7

Voltage regulators 1 —shunt

7.1 Need for voltage regulators— basic requirements

The majority of electronic circuits are designed to operate at a specified voltage, and unless the supply is at this voltage the circuit could malfunction. In addition, if the power supply has a poor regulation, a change in operating current can result in a change in input voltage. For these reasons it is often necessary to provide a voltage regulator circuit which can maintain a constant input voltage to a circuit when:

(a) there are changes in the supply voltage,
(b) there are changes in the load current.

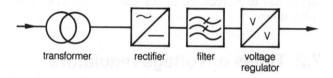

Fig. 7.1 *Block diagram of a regulated power supply*

For electronic circuits with a modest power demand, the power supply usually consists of a transformer, rectifier and filter. The transformer is usually necessary to provide a suitable input voltage because rarely would the mains voltage be exactly suitable. Even if it was, the isolation from the mains which a transformer provides is most desirable for safety reasons.

The rectifier, as we have seen in Units 3 and 4, is required to change the alternating supply to a unidirectional one, that is, ac to dc. The type of rectifier used for this purpose depends on the application and current demand of the load to be supplied, but here we will only consider very modest power demands and the use of one of the single-phase full-wave circuits.

In the smaller type power supplies we are considering, the filter is usually in the form of a single capacitor (see Unit 5) because the current demand would not require the use of anything more.

If a constant voltage is required, a voltage regulator circuit is added after the filter. This is illustrated in the block diagram in Figure 7.1, where the voltage regulator could be either a simple shunt regulator or a more complex type.

The function of the voltage regulator is to minimise the regulation (the change in voltage between no load and full load) of the rectified and filtered power supply. As we saw in Example 5.1 in Section 5, there is a fall in voltage with an increase in load and this regulation

is inherent in any rectified and filtered supply. In addition, changes in ambient temperature will also have an effect on output voltage, especially when semiconductors are concerned.

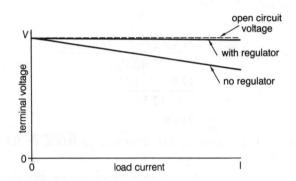

Fig. 7.2 *Voltage regulation curve of a power supply before and after fitting a voltage regulator*

No voltage regulator is perfect, and one should be chosen with the order of accuracy desired for the particular application. Figure 7.2 illustrates the regulation of a power supply before and after being fitted with a regulator. Before the regulator is fitted there is quite a large regulation and after a voltage regulator is fitted the results are very acceptable.

The definition of voltage regulation is the ratio of the difference between no load voltage and full load voltage to full load voltage. This can be written as an equation:

$$\mathrm{VR} = \frac{V_{L\,(\text{no load})} - V_{L\,(\text{full load})}}{V_{L\,(\text{full load})}} \qquad (7.1)$$

where VR is voltage regulation,

$V_{L\,(\text{no load})}$ is the no load or open circuit voltage,
$V_{L\,(\text{full load})}$ is the voltage at full load.

Example 7.1

A small dc power supply has a no load voltage of 12.6 volts. When it is loaded up to full output the load voltage falls to 11.7 volts. Determine its voltage regulation.

$$V_{L\,(\text{no load})} = 12.6\ \text{V}$$
$$V_{L\,(\text{full load})} = 11.7\ \text{V}$$
$$\mathrm{VR} = ?$$

$$\mathrm{VR} = \frac{V_{L\,(\text{no load})} - V_{L\,(\text{full load})}}{V_{L\,(\text{full load})}}$$
$$= \frac{12.6 - 11.7}{11.7}$$
$$= 0.0769$$

Answer: The power supply regulation is 0.077 or 7.7 per cent.

Example 7.2

The power supply in Example 7.1 is now fitted with a regulator and the open circuit voltage of 12.6 volts only falls to 12.5 volts at full load. What is the regulation now?

$$V_{L \text{ (no load)}} = 12.6 \text{ V}$$
$$V_{L \text{ (full load)}} = 12.5 \text{ V}$$
$$\text{VR} = ?$$

$$\text{VR} = \frac{V_{L \text{ (no load)}} - V_{L \text{ (full load)}}}{V_{L \text{ (full load)}}}$$

$$= \frac{12.6 - 12.5}{12.5}$$

$$= 0.008$$

Answer: The regulation has improved to 0.008 or 0.8 per cent.

From the two foregoing examples it can be seen the smaller the change in voltage the better the regulation.

Another function of regulation is the apparent *output resistance* of a power supply. This is similar to the internal resistance of a cell or battery which causes the terminal potential difference, tpd, of the battery to fall with an increase of current flowing from it. To determine the output resistance R_O of a power supply, we simply divide the change in voltage by the change in current which produced it. Expressing this as an equation:

$$R_O = \frac{\Delta V_L}{\Delta I_L} \tag{7.2}$$

where R_O is output resistance,
ΔV_L is change in output voltage,
ΔI_L is change in output current.

Example 7.3

From the information given in Example 7.1 determine the output resistance of the power supply if the full load current is 300 milliamperes.

$$\Delta V_L = V_{L \text{ (no load)}} - V_{L \text{ (full load)}}$$
$$= 12.6 - 11.7$$
$$= 0.9 \text{ V}$$
$$\Delta I_L = I_{L \text{ (full load)}} - I_{L \text{ (no load)}}$$
$$= 0.3 - 0$$
$$= 0.3 \text{ A}$$
$$R_O = ?$$

$$R_O = \frac{\Delta V_L}{\Delta I_L}$$

$$= \frac{0.9}{0.3}$$

$$= 3 \ \Omega$$

Answer: The output resistance of the power supply is 3 ohms.

Example 7.4

From the information given in Example 7.2, and again with a full load current of 330 milliamperes, determine the output resistance of the power supply when the voltage regulator is added.

$$\Delta V_L = V_{L \text{ (no load)}} - V_{L \text{ (full load)}}$$
$$= 12.6 - 12.5$$
$$= 0.1 \text{ V}$$
$$\Delta I_L = I_{L \text{ (full load)}} - I_{L \text{ (no load)}}$$
$$= 0.33 - 0$$
$$= 0.33 \text{ A}$$
$$R_O = ?$$

$$R_O = \frac{\Delta V_L}{\Delta I_L}$$

$$= \frac{0.1}{0.33}$$

$$= 0.30303 \ \Omega$$

Answer: The output resistance of the regulated power supply has been reduced to 0.3 ohm (an improvement of ten times).

7.2 Types of voltage regulators

There are various types of voltage regulator—from the very simple to the quite complex. Before we look at any one type in detail, we will list the different types available and briefly examine how they are able to regulate.

Shunt regulator Reference to Figure 7.3 will show the principle of operation of a shunt regulator. The load, represented by R_L, is required to have a constant voltage across it. Now this constant voltage can be provided if a constant current flows through the series resistor, R_S. This means that there will be a constant voltage drop across R_S and therefore a constant voltage across the load (and R_V in parallel with the load). Now if the load current was to vary—say fall—less current would flow through R_S, the voltage drop across it would fall and so a greater voltage will appear across the load.

It should then be apparent that if the setting of R_V was altered until the same current was again flowing through R_S, the voltage across the load would return to its original value. Now this would necessitate someone continually watching the voltage across the load and adjusting the setting of R_V so that it would remain constant.

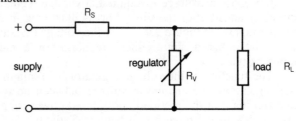

Fig. 7.3 *Principle of operation of a shunt voltage regulator*

Fortunately we have self regulating devices which will perform this function and they form the basis of the *shunt voltage regulator*.

Series voltage regulators Possibly the simplest, and oldest, method of regulating a circuit is to place a variable resistor in series with the load. Once again if the load voltage was monitored and the series resistor continually adjusted the voltage across the load could be kept constant, as in Figure 7.4. A self regulating device can be used in place of the series resistor and this then becomes a *series voltage regulator* which can maintain a constant voltage across the load.

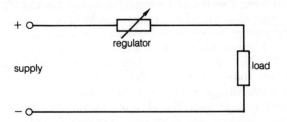

Fig. 7.4 *Principle of operation of a series voltage regulator*

Feedback series voltage regulator The self regulating series voltage regulator can be improved, if a very much closer voltage regulation is required, by the addition of a *feedback circuit*. The feedback circuit monitors the output voltage, compares this with the required voltage and corrects the error, if any. A diagrammatic representation of the system is shown in Figure 7.5, where the dashed line variable resistor symbol represents the self regulating series device.

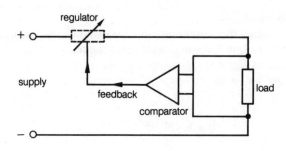

Fig. 7.5 *Principle of operation of a feedback series voltage regulator*

Linear IC voltage regulator Voltage regulators which are capable of exercising very close control become, by necessity, quite complex. To assemble a circuit that will do this task becomes rather daunting (and expensive). Fortunately by the production of *integrated circuits*, or ICs, manufacturers have been able to assemble many thousands of components in a very small physical space and at a very reasonable cost.

These *IC voltage regulators* are capable of performing the following tasks:

1. series voltage regulation with feedback;
2. over-temperature protection (150⁰C to 175⁰C);
3. over-current protection;
4. ripple rejection (typical 1000 : 1).

Used with capacitors to improve its function, a diagram of an IC voltage regulator appears in Figure 7.6.

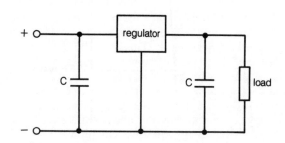

Fig. 7.6 *Representation of an IC voltage regulator*

Switch mode voltage regulator The term 'switch mode' means the rapid switching on and off of a dc supply to provide an average output. This type of control is used in modern electric traction rather than the older wasteful series resistance control.

In power supplies for electronic circuits the switched output is filtered, and this filtered level adjusted by the action of the supply circuit, to be kept constant. The average output, before filtering, is obtained by altering the 'on' time of each pulse which arrives at a constant rate. Figure 7.7 shows this effect.

Switch mode power supplies, SMPSs, are extensively employed in small demand power supplies and are almost universal in power supplies for television receivers and computers.

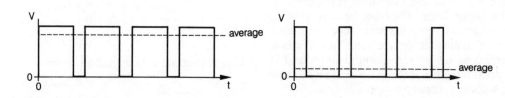

Fig. 7.7 *High (on left) and low (right) average outputs from a switched unfiltered power supply*

7.3 Zener diode shunt voltage regulator

In Section 1.10 we mentioned how the zener diode, when reverse biased, would conduct only after a certain voltage was reached. We will later discuss this effect in the *zener shunt regulator* but first let us re-examine the characteristics of the zener diode.

When forward biased the zener diode behaves like any normal PN diode in that after overcoming the barrier potential of about 0.7 volt it conducts with a normal forward resistance of a very low value. However, because of its unusual properties when reverse biased, the zener diode is only used in the reverse biased mode.

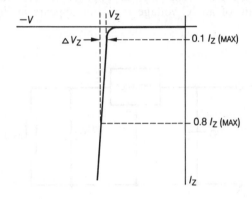

Fig. 7.8   *The reverse (operating) characteristics of a zener diode*

Figure 7.8 illustrates the reverse bias characteristics of a zener diode, always drawn in the third quandrant on a graph. On the curve there are three points of interest. The first is the point marked V_Z on the voltage axis. This point is called the zener voltage of the diode and represents the point from which the voltage across the diode (in the reverse biased condition) is considered constant. Before this point, as the voltage is increased above zero (in a negative sense on the graph), there is no conduction until, at a certain position, current commences to flow and increases until there is a linear relation between voltage and current. This transition region can be seen on the curve as a 'knee'. The point just past the 'knee' at which this linear voltage/current relationship commences, is the zener point.

From the zener point onwards is the operating mode of the zener diode. Two further positions will be noticed on this current curve—one point marked $0.1\ I_{Z(MAX)}$ and the other marked $0.8\ I_{Z(MAX)}$. These positions are respectively the minimum and maximum safe operating current for the zener diode. They are both relative to the absolute maximum current the zener diode can carry without being in danger of destruction. This absolute maximum current is set by the manufacturers and is related to the maximum power the zener diode is capable of dissipating without excessive temperature rise.

The lower current, $0.1\ I_{Z(MAX)}$, is set so that it is beyond the zener point and well into the linear part of the curve.

This minimum current is to ensure that when in operation the zener voltage will be relatively constant and will not fall below the V_Z point (i.e. the current graphed into the 'knee' of the curve).

The higher current, $0.8\ I_{Z(MAX)}$, is set so that the maximum working current is only 80 per cent of the absolute maximum, i.e. a 20 per cent margin of safety. Working between these two current points will ensure that the voltage across the zener is always stable and the zener will not be overheated and so possibly destroyed.

Zener diodes are rated firstly on their zener voltage and secondly on their maximum power ratings. The voltage ratings can vary from 2.4 volts to at least 150 volts. The power ratings for electronic circuits are usually 0.4 watt, 1 watt and 5 watts. (For power electronics use, in the industrial sphere, ratings can be very much greater than this.)

For example, a zener diode may be nominated as 8.2 volts 5 watts. This would mean, of course, that its zener voltage was 8.2 volts and the maximum power it could dissipate was 5 watts. Both these figures are used to determine the absolute maximum current and the minimum and maximum working current. Firstly consider the absolute maximum current. This may be determined from the equation:

$$I_{Z(MAX)} = \frac{P_Z}{V_Z} \qquad (7.3)$$

where P_Z is the maximum power rating of the zener diode,

V_Z is the zener voltage rating of the zener diode;

$I_{Z(MAX)}$ is the absolute maximum current of the zener diode.

Example 7.5

Determine the minimum and maximum safe working currents of an 8.2 volt 5 watt zener diode.

$$V_Z = 8.2\ \text{V}$$
$$P_Z = 5\ \text{W}$$
$$I_{Z(MAX)} = ?$$

$$\begin{aligned}
I_{Z(MAX)} &= \frac{P_Z}{V_Z} \\
&= \frac{5}{8.2} \\
&= 0.6097\ \text{A}
\end{aligned}$$

Now minimum operating current

$$\begin{aligned}
&= 0.1\ I_{Z(MAX)} \\
&= 0.1 \times 0.6097 \\
&= 0.0601\ \text{A}
\end{aligned}$$

and, maximum safe operating current

$$\begin{aligned}
&= 0.8\ I_{Z(MAX)} \\
&= 0.8 \times 0.6097 \\
&= 0.4878\ \text{A}
\end{aligned}$$

Answer: The minimum operating current of the zener diode is 60 milliamperes and the maximum safe operating current is 488 milliamperes.

A point worthy of notice is the slight slope on the linear section of the current curve in Figure 7.8. This represents resistance within the zener diode. This resistance is quite small but does produce a slight increase in voltage drop as the current increases from minimum to working maximum. This voltage increase is shown in Figure 7.8 as ΔV_Z.

To set up a zener diode as a shunt voltage regulator the circuit illustrated in Figure 7.9 is used. Note that *both* the series resistor R_S *and* the zener diode are part of the regulator circuit and both are essential for regulating purposes.

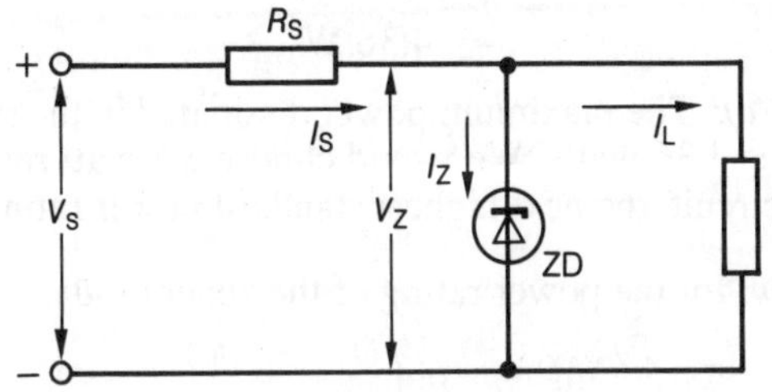

Fig. 7.9 *A zener diode shunt regulator circuit*

The voltage drop across the zener diode is constant (within limits) between the two previously discussed points on the current curve. As the zener diode is in parallel with the load, the voltage drop across the load must also be constant. Now, to keep the current through the zener diode within the two limits, the series resistor R_S must have the correct value of resistance. To determine this resistance we must know the current flowing through it and the voltage drop that must be across it.

For a start the zener diode must have at least the minimum current through it (i.e. $0.1\ I_{Z(MAX)}$) so that it is in the regulating zone or linear portion of the curve. The load current should always be taken at its maximum value. Now, the current flowing through the series resistor must be the sum of the load current and the zener diode current. This could be, from Kirchhoff's Current Law, written as an equation:

$$I_S = I_L + I_Z \qquad (7.4)$$

where I_S is the supply current through the series resistor,
I_L is the load current,
I_Z is the zener diode current.

The voltage drop across the series resistor must be the supply voltage less the zener voltage, and from Kirchhoff's Voltage Law an equation would be:

$$V_{RS} = V_S - V_Z \qquad (7.5)$$

where V_{RS} is the voltage drop across the series resistor,
V_S is the supply voltage,
V_Z is the zener voltage.

Equations 7.4 and 7.5 may be combined to give a single equation combining the two voltages and the two currents:

$$R_S = \frac{V_S - V_Z}{I_L + I_Z} \qquad (7.6)$$

where R_S is the resistance of the series resistor and the other symbols are identical with those in Equations 7.4 and 7.5.

To determine the power rating of the series resistor when the regulator is operating normally, we find the power dissipated in the resistor from the current passing through it and the voltage drop across it, as in the equation:

$$P_{RS} = V_{RS}\, I_S \qquad (7.7)$$

where P_{RS} is the power dissipated in the series resistor,
V_{RS} is voltage drop across the series resistor,
I_S is the current flowing in the series resistor.

Now if there was a possibility that the output terminals may be shorted, the complete supply voltage would appear across the series resistor. This may be very remote, and indeed impossible in some circuits, but in other circumstances provision may need to be made for it. As the full supply voltage is across the series resistor the power dissipated will be:

$$P_{RS(MAX)} = \frac{V_S^2}{R_S} \qquad (7.8)$$

where $P_{RS(MAX)}$ is the maximum power that could be dissipated in the series resistor,
V_S is the supply voltage, and
R_S is the resistance of the series resistor.

As the sum of the load current and zener current equals the supply current, and the supply current must be constant, it follows that if the load current falls to zero then the entire supply current must flow through the zener diode. As it is this current which must be no more than the maximum working current (0.8 of absolute maximum) it is essential that the power dissipated in the zener diode is known. This is the zener voltage times the sum of the minimum zener current and the maximum load current, and as an equation is written:

$$P_{Z(MAX)} = V_Z(I_Z + I_L) \qquad (7.9)$$

where $P_{Z(MAX)}$ is the maximum power that can be dissipated in the zener diode,
V_Z is the zener voltage,
I_L is the maximum load current, and
I_Z is the minimum zener diode current.

Let us now consider the application of all the above.

Example 7.6

A zener diode shunt regulator is required to reduce a 12 volt supply to 9 volts for a particular load. The load will draw a maximum current of 90 milliamperes at 9 volts. Determine the following:

(a) the resistance of the series resistor,
(b) the power rating of the series resistor under normal operation,
(c) the power rating of the series resistor if the load terminals could be shorted,
(d) the power rating of the zener diode.

$$V_S = 12 \text{ V}$$
$$V_L = 9 \text{ V}$$
$$I_L = 90 \text{ mA}$$
$$R_S = ?$$
$$P_{RS} = ?$$
$$P_Z = ?$$

From the given information we do not know the power rating of the zener diode—indeed it is a required answer to the problem. Because of this a minimum zener current must be assigned to the zener diode. This is normally set at 0.1 of the maximum load current, and if required, adjusted afterwards. Let us then assign a zener diode minimum current.

Load current is designated at 90 milliamperes, so we can assume this to be the maximum value. Therefore:

$$I_Z = 0.1 \, I_L$$
$$= 0.1 \times 0.09$$
$$= 0.009 \text{ A}$$

It is reasonable to round this figure off at 10 milliamperes as it gives an extra margin above the minimum current and simplifies calculation. We can then say the minimum zener diode current will be 10 milliamperes. Now we can proceed with the problem.

$$R_S = \frac{V_S - V_Z}{I_Z + I_L}$$
$$= \frac{12 - 9}{0.01 + 0.09}$$
$$= \frac{3}{0.1}$$
$$= 30 \, \Omega$$

Answer (a): The results of the first part of the problem suggest a resistor of 30 ohms would be suitable, and this is the correct theoretical answer. However, a 30 ohm resistor is not a standard size so for practical considerations we can select a preferred value, of the next value *below* 30 ohms—this would be 27 ohms. (For the table of preferred values of resistors see Appendix C, page 191.) We do not select a preferred value *above* 30 ohms as this may reduce the zener diode current below the minimum value.

Now determine the power ratings of the series resistor.

$$V_{RS} = V_S - V_Z$$
$$= 12 - 9$$
$$= 3 \text{ V}$$
$$I_S = I_L + I_Z$$

$$= 0.01 + 0.09$$
$$= 0.1 \text{ A}$$
$$P_{RS} = V_{RS} I_S$$
$$= 3 \times 0.1$$
$$= 0.3 \text{ W}$$

Answer (b): The resistor would need a rating of at least 0.3 watt so select from the standard rating of resistors a 0.5 watt resistor.

Now, if the load could be shorted.

$$P_{RS(MAX)} = \frac{V_S^2}{R_S}$$
$$= \frac{12 \times 12}{33}$$
$$= 4.36 \text{ W}$$

Answer (c): The maximum power dissipated in the series resistor is 4.36 watts. We would choose a 5 watt resistor for the circuit, the next highest standard power rating.

Now, for the power rating of the zener diode,

$$P_{Z(MAX)} = V_Z (I_Z + I_L)$$
$$= 9 \times (0.1 + 0.9)$$
$$= 9 \times 0.1$$
$$= 0.9 \text{ W}$$

Answer (d): The power dissipated in the zener diode is 0.9 watts. In this case we would choose a 1 watt zener diode.

Now just to check back on our first assumption of minimum zener diode current. If we use a 1 watt zener diode we can find its absolute maximum current.

$$I_{Z(MAX)} = \frac{P_Z}{V_Z}$$
$$= \frac{1}{9}$$
$$= 0.11 \text{ A}$$

The absolute maximum current would be 11 milliamperes, so 0.1 of 11 milliamperes is:

$$I_{Z(MIN)} = 0.1 \, I_{Z(MAX)}$$
$$= 0.1 \times 0.11$$
$$= 0.011 \text{ A}$$

This suggests that the minimum current should be 11 milliamperes but as we used the next lower standard value resistor below 30 ohms let us check just what this value will actually be.

$$I_S = \frac{V_S}{R_S}$$
$$= \frac{3}{27}$$
$$= 0.111 \text{ A}$$

Now we subtract the load current from this value.

$$I_Z = I_S - I_L$$
$$= 0.111 - 0.09$$
$$= 0.021 \text{ A}$$

This means that 21 milliamperes will be flowing in the zener diode which is above the minimum value required to enable it to regulate.

It can be seen from the above calculations and reasoning that as long as there is a series resistor and a zener diode in which the current flowing is above the minimum required for regulation, and the rating of the zener diode is sufficient to carry full load current, then we have a viable zener shunt regulator circuit.

Unit 7 SUMMARY

- Voltage regulators are essential if the load voltage of an electronic circuit is to remain constant.
- The ratio of the change in voltage between no load and full load to full load voltage is called the voltage regulation.
- The fall in voltage with an increase in load current is a function of the output resistance of a power supply.
- Voltage regulators improve the regulation of a power supply.
- Five types of voltage regulator circuits are:
  1. zener shunt voltage regulator
  2. series voltage regulator
  3. feedback series voltage regulator
  4. linear IC voltage regulator
  5. switched mode voltage regulator
- A zener diode operates in the reverse biased mode and we use that portion of the current curve which is linear.
- A zener shunt voltage regulator consists of a zener diode in parallel with the load, and a resistor in series with the incoming supply.

Voltage regulators 2 —series

8.1 Series voltage regulator

In Figure 7.4 we saw a series regulator in the form of a variable resistor in series with a load. This is the basis of a series regulator but is an impractical circuit because of the necessity to manually adjust the variable resistor. The principle of operation is that the voltage drop across the series resistor is subtracted from the supply voltage to give the difference as the load voltage.

In Figure 8.1 the variable resistor has been replaced with a BJT, in addition to two control devices: a resistor and zener diode. The BJT eliminates the power waste in the series resistor and performs the same task of providing a voltage drop across itself. Full load current passes through the BJT, as compared to the shunt regulator in Unit 7, so its efficiency is usually much higher. The voltage reference for the BJT series regulator is the zener diode circuit. It is the zener diode which supplies the output voltage figure, and the output voltage is always the zener voltage *less* the base-emitter voltage drop of the BJT, 0.6 volt.

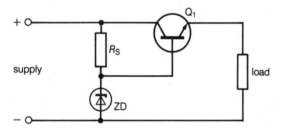

Fig. 8.1   *A series regulator circuit*

Now let us arrange the circuit of Figure 8.1 so all the voltage drops appear in vertical lines, as in Figure 8.2. First of all it can be seen that the load voltage is the difference between the supply voltage and the voltage across the BJT. This may be written as an equation:

$$V_L = V_S - V_{CE} \qquad (8.1)$$

where V_L is load voltage,
  V_S is supply voltage, and
  V_{CE} is BJT collector-emitter voltage.

Also from a study of the circuit we can see that the load is in parallel with the zener diode and the voltage drop across the base-emitter junction of the BJT. The load voltage must therefore be the difference between the zener voltage and the base-emitter voltage drop. The equation for this would be:

$$V_L = V_Z - V_{BE} \qquad (8.2)$$

where V_L is load voltage,
  V_Z is zener voltage, and
  V_{BE} is the base-emitter voltage drop of the BJT.

The voltage drop across the base-emitter junction can usually be taken as about 0.6 volt so Equation 8.2 may be written as:

$$V_L = V_Z - 0.6$$

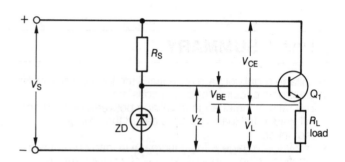

Fig. 8.2   *Voltage drops in a series regulator circuit*

Summarising the above two equations it could be said that the voltage of the load will be constant (within limits) and the difference between this voltage and the supply voltage is the BJT collector-emitter voltage which is adjusted automatically by the BJT base current.

8.2 Voltage and current relationships

In Figure 8.3 a series regulator has been once more redrawn in a more conventional layout. Some circuit and voltage values have been added. The previously quoted voltage relationships hold as before and may be restated as:

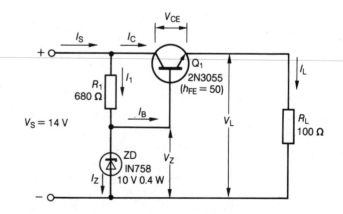

Fig. 8.3   *A practical series voltage regulator circuit showing current paths*

$$V_L = V_Z - V_{BE}$$
$$= V_S - V_{CE}$$

In addition, the current relationships may be stated as:

$$I_L = I_C + I_B$$
$$I_S = I_1 + I_C$$

but as the base current is very small (approximately $1/50$ of I_C) it is usually ignored and the equations written as:

$$I_L = I_C \qquad (8.3)$$

where I_L is the load current, and
I_C is the BJT collector current,

so that

$$I_S = I_1 + I_L \qquad (8.4)$$

In addition, the remaining current equation is:

$$I_1 = I_Z + I_B \qquad (8.5)$$

where I_1 is control circuit current,
I_Z is the zener diode current, and
I_B is the BJT base current.

Because the zener voltage is fixed, the control circuit (R_1 together with the zener diode) will maintain the collector-base voltage constant, in this case at 10 volts. The load voltage is maintained constant because the BJT, Q_1, controls the load current. The load voltage, V_L, is a function of load current (and load resistance), i.e. $V_L = I_L R_L$.

The collector current of the BJT is a function of its base current and this base current is controlled by V_{BE}. This may be shown by the equation:

$$V_{BE} = V_Z - V_L (I_L R_L)$$

Substituting with the information from Figure 8.3:

$$V_{BE} = 10 - 9.4 (94 \text{ mA} \times 100 \text{ } \Omega)$$
$$= 0.6 \text{ V}$$

The base-emitter voltage is always taken to be approximately 0.6 volt, so now we will see how this is able to control the output current and therefore the output voltage.

If the load current, for some reason, was to increase from 94 to 95 milliamperes then:

$$V_{BE} = V_Z - V_L$$
$$= 10 - (0.095 \times 100)$$
$$= 10 - 9.5$$
$$= 0.5 \text{ V}$$

Now this does not *actually* happen, as the *tendency* for V_{BE} to reduce causes the base current I_B to decrease—which, by transistor action, causes I_C to decrease and, as we have stated I_C may be taken to be I_L, so the voltage drop $I_L R_L$ (V_L) decreases. This then, by the equation last used, will return V_{BE} to its original value. This is a continuous process and irrespective of change in load current (within limits) the load voltage is maintained constant.

Let us now complete all the unknown values in the circuit of Figure 8.3.

Example 8.1

In the circuit of Figure 8.3 determine the following:
(a) load current,
(b) BJT base current,
(c) control current,
(d) zener diode current,
(e) supply current, and
(f) the voltage drop across the BJT.

$$V_S = 14 \text{ V}$$
$$V_Z = 10 \text{ V}$$
$$R_1 = 680 \text{ } \Omega$$
$$R_L = 100 \text{ } \Omega$$
$$h_{FE} = 50$$
$$I_L = ?$$
$$I_B = ?$$
$$I_1 = ?$$
$$I_Z = ?$$
$$I_S = ?$$
$$V_{CE} = ?$$

$$V_L = V_Z - V_{BE}$$
$$= 10 - 0.6$$
$$= 9.4 \text{ V}$$

$$I_L = \frac{V_L}{R_L}$$
$$= \frac{9.4}{100}$$
$$= 0.094 \text{ A}$$

Answer (a): The load current is 94 milliamperes.

$$I_B = \frac{I_C}{h_{FE}}$$
$$= \frac{0.094}{50}$$
$$= 1.88 \times 10^{-3} \text{ A}$$

Answer (b): The BJT base current is 1.88 milliamperes.

$$I_1 = \frac{V_S - V_Z}{R_1}$$
$$= \frac{14 - 10}{680}$$
$$= \frac{4}{680}$$
$$= 5.88 \times 10^{-3} \text{ A}$$

Answer (c): The control current is 5.88 milliamperes.

$$
\begin{aligned}
I_Z &= I_1 - I_B \\
&= 5.88 \times 10^{-3} - 1.88 \times 10^{-3} \\
&= 4 \times 10^{-3} \text{ A}
\end{aligned}
$$

Answer (d): The zener diode current is 4 milliamperes.

$$
\begin{aligned}
I_S &= I_C + I_1 \\
&= 94 \times 10^{-3} + 5.88 \times 10^{-3} \\
&= 99.88 \times 10^{-3} \text{ A}
\end{aligned}
$$

Answer (e): The supply current is 100 milliamperes.

$$
\begin{aligned}
V_{CE} &= V_S - V_L \\
&= 14 - 9.4 \\
&= 4.6 \text{ V}
\end{aligned}
$$

Answer (f): The BJT collector-emitter voltage drop is 4.6 volts.

The student should confirm that all of the above is correct, and check by putting all the figures given into the equations so far discussed. Check also that the zener diode current is not less than the minimum current for regulation.

8.3 The Darlington pair series voltage regulator

Most power BJTs, as used in series regulators, are capable of carrying relatively high currents but do not have a high current gain (h_{FE}). This means that the base current, although admittedly quite low, may slightly change the operating voltage of the zener diode (remember the zener voltage is not absolutely constant and does increase with increase in zener diode current). By using a circuit called a *Darlington pair* it is possible to have a very high current gain and thereby keep base current very low.

The circuit in Figure 8.4 is similar to that in Figure 8.3 with the addition of an extra BJT, Q_1. When connected in this fashion the two BJTs operate as if they were a single unit. Note that the collectors of both BJTs are connected together and form a common external lead,

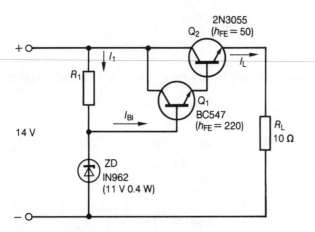

Fig. 8.4 *A Darlington pair series voltage regulator circuit*

the emitter of Q_1 connects to the base of Q_2 as an internal connection, the base of Q_1 forms an external base lead and the emitter of Q_2 is an external emitter. In fact it is possible to buy Darlington BJTs as a single unit with only three external leads. This fact was mentioned in Section 6.9, and Figure 6.11 illustrates a Darlington BJT, designated BD263. However here we use two separate units to illustrate the fact that a Darlington BJT is actually made from two BJTs.

The total current gain of a Darlington pair is the product of the individual gains of each BJT. This is written as:

$$
h_{FE}(\text{total}) = h_{FE1} \times h_{FE2} \tag{8.6}
$$

As we can also use the symbol β (Greek letter beta) for BJT current gain, Darlington pairs are sometimes referred to as β multiplier circuits.

The operation of a Darlington series voltage regulator is little different than when using a single BJT except for two features. The first is that the base-emitter voltage drops of the two BJTs must be added, as they are effectively in series. As this voltage drop is usually taken as 0.6 volt, the Darlington pair has a combined V_{BE} voltage drop of 1.2 volts. In the series regulator circuit this must be subtracted from the zener voltage to obtain the load voltage. The second difference is the very low base current. Let us examine the operating conditions of a Darlington series regulator.

Example 8.2

For the circuit of Figure 8.4, determine the following:

(a) load voltage,
(b) load current,
(c) base current I_{B1},
(d) zener diode current,
(e) control current (current flowing in R_1),
(f) resistance of R_1, and
(g) power dissipated in R_1.

$$
\begin{aligned}
V_S &= 14 \text{ V} \\
V_Z &= 11 \text{ V} \\
h_{FE1} &= 220 \\
h_{FE2} &= 50 \\
R_L &= 10 \text{ }\Omega \\
V_L &= ? \\
I_L &= ? \\
I_{B1} &= ? \\
I_Z &= ? \\
I_1 &= ? \\
R_1 &= ? \\
P_{R1} &= ?
\end{aligned}
$$

$$
\begin{aligned}
V_L &= V_Z - (V_{BE1} + V_{BE2}) \\
&= 11 - (0.6 + 0.6) \\
&= 9.8 \text{ V}
\end{aligned}
$$

Answer (a): The load voltage is 9.8 volts.

$$I_L = \frac{V_L}{R_L}$$

$$= \frac{9.8}{10}$$

$$= 0.98 \text{ A}$$

Answer (b): The load current is 980 milliamperes.

$$I_{B1} = \frac{I_L}{h_{FE1} \times h_{FE2}}$$

$$= \frac{0.89}{220 \times 50}$$

$$= 89 \times 10^{-6} \text{ A}$$

Answer (c): The base current of Q_1 is 89 microamperes.

$$I_{Z(MAX)} = \frac{P_{Z(MAX)}}{V_Z}$$

$$= \frac{0.4}{11}$$

$$= 0.036 \text{ A}$$

The maximum zener diode current is 36 milliamperes. To find minimum zener diode current,

$$I_{Z(MIN)} = 0.1 \, I_{Z(MAX)}$$

$$= 0.1 \times 0.036$$

$$= 0.0036 \text{ A}$$

Answer (d): The minimum zener diode current is 3.6 milliamperes.

This is a *very* small current and since it is the minimum that should flow through the zener diode for reliable regulation it is preferable to increase it to, say, 5 milliamperes for a more workable current. Now, to find control current,

$$I_1 = I_Z + I_{BE1}$$

$$= 5 \times 10^{-3} + 89 \times 10^{-6}$$

$$= 5.089 \times 10^{-3} \text{ A}$$

Answer (e): The *very* small base current may be ignored and we can say I_1 is 5 milliamperes.

$$R_1 = \frac{V_S - V_Z}{I_1}$$

$$= \frac{14 - 11}{5 \times 10^{-3}}$$

$$= 600 \, \Omega$$

Answer (f): The calculated resistance is 600 ohms so use the next lower preferred value, 560 ohms.

$$P_{R1} = I_{R1} (V_S - V_Z)$$

$$= 0.005 \times (14 - 11)$$

$$= 0.015 \text{ W}$$

Answer (g): The power dissipated in the resistor R_1 is only 15 milliwatts. As the lowest rated resistor in common

use is 250 milliwatts, this dissipation is well within its rating, so we would use a ¼ watt resistor.

8.4 Feedback series voltage regulator

The Darlington series regulator, as discussed in Section 8.3, can be further improved in its regulating ability by the use of a feedback amplifier. The feedback is achieved by monitoring the voltage across the load. A change in this voltage changes the base current of Q_3, which adjusts the V_{CE} voltage drop across the series BJT, Q_2. This change in the V_{CE} voltage of Q_2 is such that the load voltage is returned to its original value, and therefore held constant.

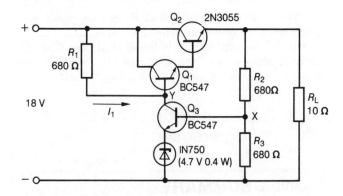

Fig. 8.5 *A feedback Darlington series voltage regulator circuit*

In the circuit of Figure 8.5, which is a feedback series regulator, a proportion of the output voltage at the point x is fed to the base of BJT Q_3. The point x is formed by the voltage divider action of R_2 and R_3. The collector current, and therefore the V_{CE} voltage across Q_3, controls the base current of Q_1 of the Darlington pair. This in turn, by transistor action, will change the V_{CE} voltage drop across Q_2 and in turn change the load current which will restore the voltage at the load to its former value.

As the load voltage across R_L is also across the series combination of R_2 and R_3, these two voltages must be equal. The voltage at point x must always be the zener voltage plus the base-emitter voltage of Q_3. Therefore the output voltage must equal this voltage times the voltage divider effect of R_2 and R_3. This may be written as:

$$V_L = \frac{(V_Z + V_{BE}) \times (R_2 + R_3)}{R_3} \quad \text{(8.7)}$$

where V_Z is the zener voltage,
V_{BE} is 0.6 volt,
R_2 is the upper voltage divider resistor,
R_3 is the lower voltage divider resistor,
V_L is the load voltage.

Example 8.3

From the values given in the circuit of Figure 8.5, determine the voltage at the load.

$$V_Z = 4.7 \text{ V}$$
$$R_1 = 680 \ \Omega$$
$$R_2 = 680 \ \Omega$$
$$R_3 = 680 \ \Omega$$
$$V_L = \ ?$$

$$
V_L = \frac{(V_Z + V_{BE}) \times (R_2 + R_3)}{R_3}
$$

$$
= \frac{(4.7 + 0.6) \times (680 + 680)}{680}
$$

$$
= 10.6 \text{ V}
$$

Answer: The load voltage is 10.6 volts.

The combination of R_2 and R_3 can be adjusted to give a variation above and below the voltage as calculated above. If the combination is made as a potentiometer variable resistor, a voltage between $(V_Z + V_{BE})$ and a few volts below the supply voltage may be selected.

The voltage drop across R_1, $I_1 R_1$, governs the base current to the base of Q_1, I_{B1}. Now, if the voltage across the load tends to rise, the voltage at the base of Q_3 rises. This increases the base current of Q_3, and by transistor action increases the collector current of Q_3. This increased current in Q_3, I_Z, also increases I_1. This in turn increases the voltage drop across R_1, which reduces the voltage at the base of the Darlington pair, point y. Less base current will flow into the Darlington pair and so, by transistor action, the collector current (which may be considered load current) is reduced and the voltage returns to its original value, as:

$$V_L = I_L R_L$$

Even if the load current changes, the voltage will be held constant by the procedure outlined above.

The resistance of resistor R_1 is determined by taking the current through it as about 0.2 of the maximum current of the zener diode, and the voltage across it as the supply voltage less the zener voltage. The values of the voltage divider resistors, R_2 and R_3, are made to total twice the resistance of R_1.

Unit 8 SUMMARY

- The series voltage regulator is usually more efficient than the shunt type.
- The load voltage in a series voltage regulator is determined by the voltage of the control zener diode less the BJT base-emitter voltage drop.
- A Darlington pair uses two BJTs in cascade to produce a very high current gain combination capable of carrying relatively high currents.
- A series voltage regulator, using a Darlington pair, is capable of better voltage regulation than when using a single power BJT.

- Even better voltage regulation may be obtained by using a feedback circuit and an extra BJT.
- The feedback series voltage regulator can be made into an adjustable voltage regulator by using a potentiometer connected variable resistor in place of the fixed voltage divider.

Unit 9

Voltage regulators 3 —IC regulators

9.1 Fixed output IC voltage regulator

By modern production techniques it has been possible to produce quite sophisticated voltage regulators as integrated circuits (or ICs) at comparatively very low cost. This was mentioned in Section 7.1 and the products range from the simpler three terminal fixed voltage types to more expensive precision regulators.

The fixed voltage regulators are usually made to three standard voltages: 5 volt, 12 volt and 15 volt, although others may be available. They have many features built into them which make them very desirable as reliable, safe, and easy-to-install regulators. Since many circuits have either a positive or negative polarity as reference to earth, or a common point, three-terminal IC regulators may be designated either positive or negative. One popular regulator is designated 78xx if positive and 79xx if negative. (The 'x' after the first two digits is the fixed output voltage of the regular, so one designated 7812 is a 12 volt positive regulator.) Both positive, 7815, and negative, 7915, three-terminal regulators can be seen in Figure 9.1. Additionally, regulators are available as dual polarity regulators: these present both a positive and negative regulated voltage, with reference to a common point.

Whilst the majority of regulators in use have a maximum current rating of 1 to 1.5 amperes, they are also available with current ratings of up to 10 amperes.

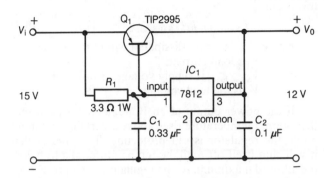

Fig. 9.2 *A current boost IC voltage regulator circuit*

Additionally, they may be used with an external power BJT to extend their current ratings. These are sometimes termed *current-boost* circuits and one is illustrated in Figure 9.2.

While three-terminal IC regulators will present a constant (within limits) output voltage, to do this the input voltage must be greater than, but not too much greater than, the nominal output voltage. The minimum difference between input and output voltage varies from 2 volts to 3 volts, depending on output voltage and type of regulator. It is usually considered that the difference should be no less than 3 volts. This is sometimes called the *dropout voltage* because the regulator will not function, or 'drops out' of regulation, if the input falls within this figure. Another term for this minimum differential is the *regulator headroom*. The maximum voltage at which the regulator will safely operate without suffering damage is typically 35 volts.

The power dissipated within an IC regulator device is taken as the current being passed, times the difference

Fig. 9.1 *Positive and negative IC voltage regulators (circled) in an amplifier constructed by the author*

between input and output voltage. Written as an equation, this becomes:

$$P_{(tot)} = I_L (V_{S\,(full\,load)} - V_L) \quad (9.1)$$

where $P_{(tot)}$ is power dissipated within the device,
  I_L is load current,
  $V_{S(full\,load)}$ is supply voltage at full load,
  V_L is the load voltage.

If the power output of an IC regulator is such that the internal temperature rises above a certain maximum point, the regulator is self protecting. This means that it will shut itself off until the temperature falls to below the allowed maximum. As the regulator is self protecting, it follows that to allow it to dissipate the maximum amount of heat without overheating it should be attached to a heatsink which will dissipate heat to the surrounding air. An IC regulator operating well below its maximum rating would not require the heat removal abilities of one operating at full current rating.

In addition to the over power shut down abilities, the same internal circuitry will shut down the regulator if the ambient temperature plus the IC regulator's own temperature rise exceeds the maximum allowed value. In this case the temperature of the surroundings could limit the maximum current the IC regulator could carry without shutting itself down.

An inbuilt current sensor limits the current to the maximum current the device is designed to carry. If the current demand exceeds this limit, the output voltage will fall while maintaining that current.

One very great advantage in using an IC regulator is its ripple rejection ability because ripple is just a small regular variation in input voltage. A typical IC regulator would reduce the ripple content of a rectified and filtered power supply to as low as 3 per cent of its input value. So the IC regulator supplies an added bonus in being a second, and very effective, filter. IC regulators come in different package styles. Figure 9.3 illustrates two common packages.

Unregulated power supplies usually have a poor voltage regulation and the regulator is added to greatly improve this. However before a regulator can be added the conditions of drop out must be adhered to. Most power supplies are designed for a particular output voltage at no load, and if filter capacitors are used this will be the peak, or maximum, value of the input waveform. When loaded this output voltage falls and this lower voltage is the one that must be considered for drop out. Let us determine the no load voltage for a rectified and filtered power supply for use with a regulator.

Example 9.1

A power supply is known to have a voltage regulation of 25 per cent and is to be used with a 12 volt IC regulator, for a 12 volt supply to a load. What no load voltage must the power supply have so that at full load it will be above the drop-out voltage?

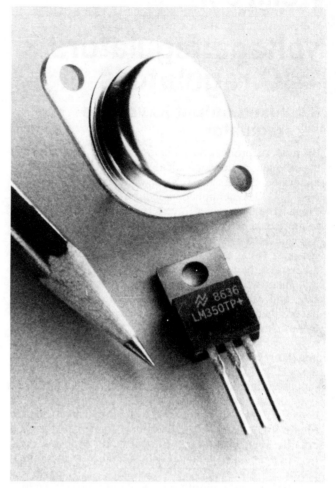

Fig. 9.3 *Two typical three-terminal IC voltage regulators*

$$V_R = 25 \text{ per cent}$$
$$V_L = 12 \text{ V}$$
$$V_{(drop\,out)} = 3 \text{ V (understood)}$$
$$V_{(no\,load)} = ?$$

The output voltage of the regulator (V_{reg}) is the same as the load voltage (V_L) so the minimum input voltage to the IC regulator must be the load voltage plus the drop-out voltage ($V_{(drop\,out)}$). This minimum input voltage would be at the worst condition, full load. To find this minimum input voltage ($V_{(full\,load)}$) we can make up the equation:

$$V_{(full\,load)} = V_L + V_{(drop\,out)}$$
$$= 12 + 3$$
$$= 15 \text{ V}$$

Now, transposing Equation 7.1

$$V_{(no\,load)} = (V_{(full\,load)} \times VR) + V_{(full\,load)}$$
$$= (15 \times 0.25) + 15$$
$$= 18.75 \text{ V}$$

Answer: The no load voltage of the power supply must be at least 18.8 volts.

When using an IC voltage regulator in a practical circuit various precautions are usually taken to ensure reliable operation and also to protect the IC. Referring to Figure 9.4 it can be seen there are two capacitors and a diode added to the IC in the depicted circuit. The first capacitor, C_1, is needed if the regulator is located any distance from the power supply filter capacitor. Unless the IC is mounted right beside the filter capacitor this is usually included. Capacitor C_2 is added to improve transient response of the regulator, i.e. its ability to act quickly to correct any tendency to change the output voltage. The diode is necessary if there is any chance of the output of the power supply being short circuited. If this happened the charged capacitor C_2 would, in effect, be in parallel with the IC and would place across it the full voltage in the reverse direction. This could result in the destruction of the IC. With the diode in the circuit the only voltage across the IC would be the forward voltage drop of the diode (0.6 volt), which could cause no damage whatsoever.

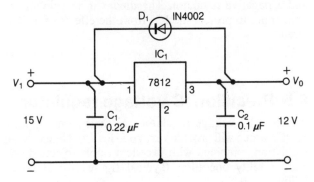

Fig. 9.4 *An IC voltage regulator circuit with external protective components*

9.2 The current boost circuit

The current boost circuit is illustrated in Figure 9.2. In this circuit the IC voltage regulator is acting as a control, and the voltage regulation is carried out by the series power BJT. The IC regulator shown in the circuit is capable of delivering a little over one ampere. If the current demand of the load is higher than this the BJT takes over, still controlled by the IC. However it is usual for the BJT to take over at a much lower current.

The resistance of resistor R_1 controls the point when the BJT commences conducting and sharing the load current. This resistor is often termed a *current sensing* resistor. When the load commences to draw current the IC supplies this through resistor R_1. As this current increases, the voltage drop across R_1 increases until it reaches 0.6 volt. At this point this voltage forward biases the BJT and it commences to conduct. The resistor designated R_1, in the circuit of Figure 9.2, is sometimes given the designation R_{SC} and this will be used in the calculation to follow. The current the IC will conduct

before the BJT assists may be determined from the Ohm's Law equation:

$$I_{R1} = \frac{V_{BE}}{R_{SC}}$$

where I_{R1} is the current through R_{SC},
V_{BE} is the base-emitter voltage drop, and
R_{SC} is the current sensing resistor.

In the circuit of Figure 9.2 we have the following values:

$$V_{BE} = 0.6 \text{ V (understood)}$$
$$R_{SC} (R_1) = 3.3 \text{ } \Omega$$
$$I_{RSC} = ?$$

$$I_{RSC} = \frac{0.6}{3.3}$$
$$= 0.1818 \text{ A}$$

So the IC would pass up to 182 milliamperes before the BJT would assist it. As the current increases further, the voltage divider action of the IC and the current sensing resistor (R_1) sets the BJT base current and, by transistor action, the collector (load) current, which maintains a constant voltage drop across the load. In the circuit depicted in Figure 9.2 the output current can be a little in excess of 10 amperes.

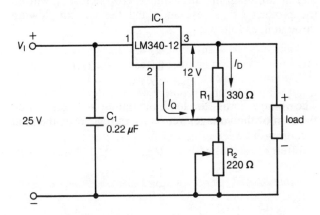

Fig. 9.5 *External resistors added to provide a variable output from an IC voltage regulator*

9.3 A variable output from an IC voltage regulator

If an IC voltage regulator is arranged in a circuit, such as in Figure 9.5, it may be used as a variable-voltage IC voltage regulator. Note that the common terminal (terminal 2) now connects to the junction of a voltage divider network, formed by R_1 and R_2. The output voltage may be varied upwards from the nominal output voltage of the IC. There are two current paths in the voltage divider: one from the IC common terminal, marked I_Q in the diagram, and the nominal voltage-divider current through R_1, marked I_D.

In the depicted circuit (Fig. 9.5) the output voltage of the IC voltage regulator is 12 volts. This is in parallel with R_1 so the voltage drop across R_1 must always be 12 volts. With the value of R_1 shown in the circuit (330 ohms), by Ohm's Law the current through R_1 (I_D) would be:

$$I_D = \frac{V_L}{R_1}$$
$$= \frac{12}{330}$$
$$= 0.0364 \text{ A}$$

So a current of 36 milliamperes will flow through R_1. Now the current to operate the regulator (I_Q), often called the IC *quiescent current*, is about 8 milliamperes for this type of regulator, so the current through resistor R_2 will be the sum of these two currents, as follows:

$$I_{R_2} = I_D + I_Q$$
$$= 0.0364 + 0.008$$
$$= 0.0444 \text{ A}$$

So the current through variable resistor R_2 will be approximately 44 milliamperes.

Now if the variable resistor, R_2, is adjusted so that it is at maximum resistance (variable contact at the lower end in the diagram) the voltage drop across it will be the product of its resistance and the current flowing through it, as follows:

$$V_{R_2} = I_{R_2} R_2$$
$$= 0.0444 \times 220$$
$$= 9.67 \text{ V}$$

The voltage across the load, the output voltage, will be the sum of the voltages across R_1 and R_2. Thus the load voltage will be:

$$V_L = V_{RS1} + V_{RS2}$$
$$= 12 + 9.67$$
$$= 21.7 \text{ V}$$

Therefore the voltage is able to be varied from 12 volts to about 21 volts. Wherever it is set, by moving the R_2 contact from the upper end to the lower end in the diagram, it will remain substantially constant over the current limits and input voltage limits of the regulator. However the regulating ability of this circuit is inferior to that of an IC used as a fixed voltage regulator since the quiescent current, I_Q, changes slightly with load current. This in turn will increase the voltage drop across R_2 and cause a slightly lower voltage at the output. A larger range of voltage can be obtained by increasing the value of R_2 but at the expense of even more inferior regulation.

9.4 A dual-polarity regulator circuit

A dual-polarity (or bipolar) regulator is shown in Figure 9.6. In this circuit, use has been made of both a positive and a negative regulator. The diodes in the circuit are once more to protect the ICs from the effects of a short circuit.

9.5 Precision IC voltage regulator

Precision IC voltage regulators, as their name implies, are ICs which will provide very good load and line voltage regulation combined with excellent ripple rejection and high stability operation. Typical figures would be as follows:

- load regulation better than 0.1 per cent
- dc line regulation 0.03 per cent per volt
- ripple rejection 0.01 per cent per volt
- output voltage adjustable from 4.5 to 40 volts
- output current 45 milliamperes, which may be extended to in excess of 10 amperes by external BJTs

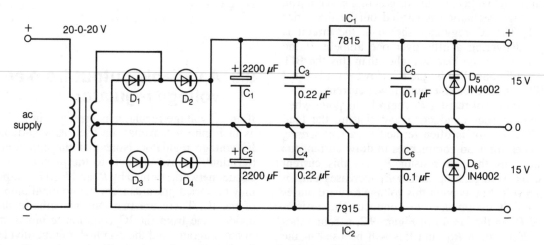

Fig. 9.6 *A positive and negative IC voltage regulator arranged to give a dual polarity voltage supply*

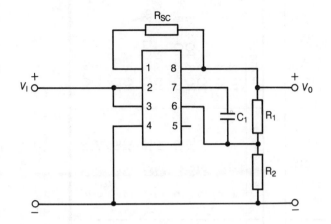

Fig. 9.7 *Basic circuit for an IC precision voltage regulator*

A typical simple circuit using a precision IC voltage regulator is shown in Figure 9.7. The output voltage is adjustable by the selection of the resistance of the two resistors, R_1 and R_2. In addition, current limiting is provided by the current sensing resistor R_{SC}. The values of the circuit components may be selected by using the following approximate equations:

$$R_1 \simeq 1.11 \times 10^3\, V_{(out)}$$

$$R_2 \simeq \frac{1.72\, R_1}{V_{(out)} - 1.72}$$

$$R_{SC} \simeq \frac{325 \times 10^{-3}}{I_{SC}}$$

where $V_{(out)}$ is the regulated output voltage,
   R_1 is the upper resistor in the circuit of Figure 9.6,
   R_2 is the lower resistor in the circuit of Figure 9.6,
   R_{SC} is the current sensing resistor,
   I_{SC} is the sensing current.

(It is not suggested that the student should memorise these approximate equations and they are included here for interest only.)

Figure 9.8 shows a circuit, using a precision IC voltage regulator, with an output adjusted (with 1 per cent precision resistors) to 28 volts and with an output current capacity of one ampere. The diodes are included for circuit protection. D_1 protects against output voltage reversal, D_2 against shorted input and D_3 against input voltage reversal. Note that the series boost BJT, Q_2, is driven by the switching BJT, Q_1.

Figure 9.9 shows a precision regulator circuit which includes *foldback current limiting*. This particular circuit is adjusted for an output voltage of 15 volts and limited to an output current of 200 milliamperes. If the current exceeds this value the circuit will not only prevent further current from flowing but will reduce the current to a value which will not produce excessive power loss in the IC. This effect may be seen in the accompanying graph, where the foldover current is reduced to about one third of its maximum value.

Table 9.1 lists the voltage regulators available from one manufacturer at the time of printing. Readers may see the range of current outputs, voltage outputs and packages from which a required regulator could be chosen.

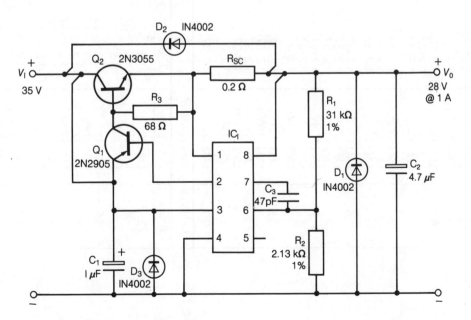

Fig. 9.8 *A precision IC voltage regulator current boost circuit*

Table 9.1 Voltage regulator selection guide

1. Adjustable regulators (a) Positive current output

Amps	Device	Output voltage	Package
10	LM196	1.25–15	TO–3
	LM396	1.25–15	TO–3
5	LM138	1.2–32	TO–3
	LM238	1.2–32	TO–3
	LM338	1.2–32	TO–3
3	LM150	1.2–33	TO–3
	LM250	1.2–33	TO–3
	LM350	1.2–33	TO–3, TO–220
1.5	LM117	1.2–37	TO–3
	LM117HV	1.2–57	TO–3
	LM217	1.2–37	TO–3
	LM217HV	1.2–57	TO–3
	LM317	1.2–37	TO–3, TO–220
	LM317HV	1.2–57	TO–3
0.5	LM117H	1.2–37	TO–39
	LM117HVH	1.2–57	TO–39
	LM217H	1.2–37	TO–39
	LM317H	1.2–37	TO–39
	LM317HVH	1.2–37	TO–39
	LM317M	1.2–57	TO–202
0.1	LM317L	1.2–37	TO–92

(b) Negative current output

Amps	Device	Output voltage	Package
3	LM133	–1.2 – –27	TO–3
	LM333	–1.2 – –27	TO–3, TO–220
	LM333A	–1.2 – –27	TO–3
1.5	LM137	–1.2 – –37	TO–3
	LM137HV	–1.2 – –47	TO–3
	LM237	–1.2 – –37	TO–3
	LM237HV	–1.2 – –47	TO–3
	LM337	–1.2 – –37	TO–3, TO–220
	LM337HV	–1.2 – –47	TO–3
0.5	LM137H	–1.2 – –37	TO–39
	LM137HVH	–1.2 – –47	TO–39
	LM237H	–1.2 – –37	TO–39
	LM337H	–1.2 – –37	TO–39
	LM337HVH	–1.2 – –47	TO–39
	LM337M	–1.2 – –37	TO–202
0.1	LM337L	–1.2 – –37	TO–92

2. Fixed regulators (a) Positive current output

Amps	Device	Output voltage	Package
3	LM123	5	TO–3
	LM223	5	TO–3
	LM323	5	TO–3
1.5	LM109	5	TO–3
	LM209	5	TO–3
	LM309	5	TO–3
	LM140	5, 12, 15	TO–3
	LM140A	5, 12, 15	TO–3
	LM340	5, 12, 15	TO–3, TO–220
	LM340A	5, 12, 15	TO–3, TO–220
	LM78XXC	5, 12, 15	TO–3, TO–220
0.5	LM341	5, 12, 15	TO–220, TO–202
	LM78MXXC	5, 12, 15	TO–220
0.2	LM109H	5	TO–39
	LM209H	5	TO–39
	LM309H	5	TO–39
	LM342	5, 12, 15	TO–202
0.1	LM140LA	5, 12, 15	TO–39
	LM340L	5, 12, 15	TO–92, TO–39
	LM78LXXA	5, 12, 15	TO–92, TO–39

(b) Negative current output

Amps	Device	Output voltage	Package
3	LM145	–5, –5.2	TO–3
	LM345	–5, –5.2	TO–3
1.5	LM120	–5, –12, –15	TO–3, TO–220
	LM320*	–5, –12, –15	TO–3, TO–220
	LM79XXC	–5, –12, –15	TO–3, TO–220
0.5	LM320M*	–5, –12, –15	TO–202
	LM79MXXC	–5, –12, –15	TO–202, TO–39
0.2	LM120H	–5, –12, –15	TO–39
	LM320H	–5, –12, –15	TO–39
0.1	LM320L*	–5, –12, –15	TO–92
	LM79LXXAC	–5, –12, –15	TO–92

Temperature operating range: LM100 Series –55°C to +150°C; LM200 Series –25°C to +125°C; LM300 Series 0°C to +125°C;   COURTESY NATIONAL SEMICONDUCTOR

* The LM320 has better electrical characteristics than the LM79XX.

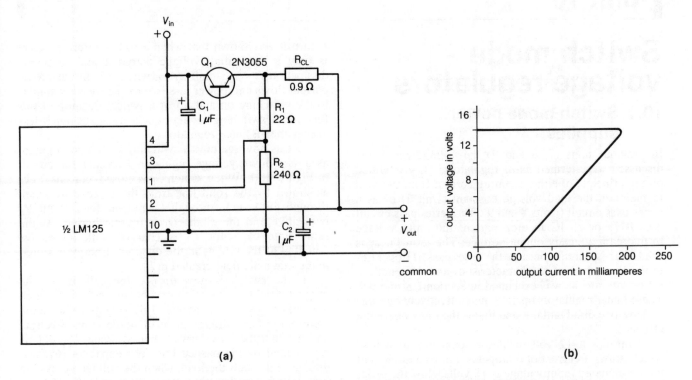

(a)   **(b)**

Fig. 9.9 *(a) A precision IC voltage regulator circuit providing foldback current limiting and (b) graph showing the current foldback after a short circuit*

Unit 9 **SUMMARY**

- IC voltage regulators are single devices which are capable of very good voltage regulation.
- Three terminal IC voltage regulators have input, output and common terminals.
- IC voltage regulators are available as positive, negative and dual polarity types.
- Three standard voltages for IC voltage regulators are 5, 12, and 15 volts.
- The output current capability of an IC voltage regulator may be extended with external BJTs. This is then termed a current boost circuit.
- The minimum difference between input and output voltage of an IC voltage regulator must be no less than 3 volts. This is usually referred to as drop out voltage.

- An IC voltage regulator will shut itself down if its temperature rises above a maximum value and will limit its output current to a set maximum value.
- By using external resistors the output voltage from a fixed voltage regulator may be made variable above the nominal output voltage.
- Precision IC voltage regulators give excellent regulation and ripple rejection. They may be used with external components for such features as current boosting and current foldover. Their output voltage may be set to any value within their capability by external resistors.

Unit 10

Switch mode voltage regulators

10.1 Switch mode power supplies

In previous units (7, 8 and 9) the voltage regulators discussed were termed *linear* regulators. They operated on the principle of either shunting current from the load to maintain the load voltage constant (Unit 7) or as a series pass circuit (Units 8 and 9). The series pass circuit uses BJTs or an IC to drop part of the input voltage to maintain a steady output voltage. The term linear is used for these circuits because the series pass BJT operates in the linear part of its characteristics and does not cut off or saturate, as was explained in Section 6.8 and 6.9. These linear regulators operate quite effectively but they do have one disadvantage, and that is their low operating efficiency.

Consider a 12 volt regulator connected to a load which draws a current of 6 amperes. Let us assume that the unregulated input voltage is 18 volts. Now the series pass BJT (or the inbuilt BJT in the IC) has to drop part of this voltage as follows:

$$V_{(drop)} = V_{(supply)} - V_{(load)}$$
$$= 18 - 12$$
$$= 6 \text{ V}$$

So 6 volts is dropped across the series device, and the current flowing through it (the load current) is 6 amperes. Now the power dissipated in the regulator may be determined by using Equation 9.1.

$$P_{(tot)} = I_S (V_S - V_L)$$
$$= 6 \times (18 - 12)$$
$$= 36 \text{ W}$$

This power will require quite a large heatsink to dissipate it to the surrounding air and keep the temperature within limits.

Let us consider the efficiency of the regulator. The efficiency equation can be written as:

$$\eta = \frac{V_S I_L}{V_L I_L}$$
$$= \frac{12 \times 6}{18 \times 6}$$
$$= 0.6666$$

The efficiency is only 66.7 per cent and when we consider losses in the supply transformer and rectifier diodes it can be seen that linear regulated power supplies could have efficiencies of less than 50 per cent.

A much higher efficiency may be achieved by replacing the linear regulator with a switching BJT. In

Unit 6 it was shown that when a BJT is saturated there is only a very small voltage across it and so power dissipation is very small. Also, when it is in cut off no current flows and power loss is zero. This is very similar to the switching on and off of a switch. Because of this function a switching regulator will dissipate much less energy than a linear regulator.

Let us now see how a switching circuit can operate as a step-down voltage regulator. Consider the circuit in Figure 10.1. When switch S_1 is closed, current flows as shown by the solid line from the positive terminal, through the switch, through L_1 and the load. Point V_X is positive and the diode, D_1 is reverse biased. At the same time capacitor C_1 is charged with a positive potential at the top, as in the diagram. Energy is stored in the magnetic field created in L_1.

If the switch is now opened the collapse of the magnetic field produces an emf of self induction which will make point V_X negative. This potential will allow a current to flow through the load, in the same direction, and through the now forward biased diode, D_1. This is represented by the dashed line. The capacitor, C_1, will discharge through the load, when the voltage across the load falls below the capacitor voltage, which produces a relatively steady dc output.

When the switch is closed, once more current will flow through the load in the same direction, limited by the now opposite emf of self induction in L_1 produced by the rising current. If this sequence is repeated and the switch is closed longer than it is open, the voltage will be higher than if the switch is open longer than it is closed.

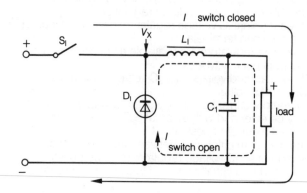

Fig. 10.1 *The action of a switch mode power supply*

By differing circuit configurations, switch mode power supplies can be made step-up, step-down and inverting (i.e. change of polarity with respect to a common point). In addition, if the switching frequency is high they operate at very high efficiencies (above 80 per cent) and the difference in input and output voltage has little effect on efficiency.

10.2 Basic step-down switching regulator

In Figure 10.2 the BJT, Q_1, has replaced the switch of the circuit in Figure 10.1. Below Q_1, connected to its base, is a block diagram designated 'pulse width modulator'. The action of the pulse width modulator is to supply the 'on' and 'off' pulses to the BJT. When Q_1 is supplied with the optimum base current it switches on, it saturates, and the voltage drop across it is negligible and full current flows. When the base current is suddenly removed it switches off and current ceases to flow.

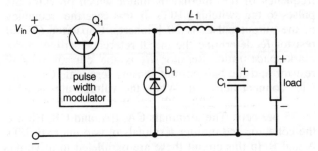

Fig. 10.2 *A basic step-down switch mode voltage regulator*

The action of pulse width modulation may be seen in Figure 10.3. The upper diagram shows the current flowing in the load, as a constant value, and the rising and falling current through the inductor. The lower diagram shows the on and off switching of the BJT and the mark/space ratio. T is the time for one cycle of switching. The average output voltage may be determined from the mark/space ratio times the input voltage. This may be written as an equation:

$$V_{(OUT)} = V_{(IN)}\frac{t_{(ON)}}{T}$$

where $V_{(OUT)}$ is the output (load) voltage,
$V_{(IN)}$ is the input voltage,
$t_{(ON)}$ is the time in a cycle the BJT is conducting,
T is the time of the complete cycle.

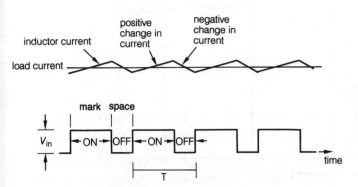

Fig. 10.3 *Mark/space switching voltage, and load and inductor currents, in a switch mode circuit*

Example 10.1

Determine the output voltage of a switch mode voltage regulator with an input voltage of 15 volts and operating at a frequency of 20 kilohertz. The mark/space ratio is 2:1.

$$V_{(IN)} = 15 \text{ V}$$
$$f = 20 \text{ kHz}$$
$$\text{mark/space ratio} = 2:1$$
$$V_{(OUT)} = ?$$

$$T = \frac{1}{f}$$
$$= \frac{1}{20 \times 10^3}$$
$$= 50 \times 10^{-6} \text{ s}$$

Periodic time is 50 microseconds, and as $T = t_{ON} + t_{OFF}$, and as the mark/space ratio is 2:1, the 'on' time is 2/3 of 50 microseconds, which is 33.3 microseconds. Now,

$$V_{(OUT)} = V_{(IN)}\frac{t_{(ON)}}{T}$$
$$= 15 \times \frac{33.3}{50}$$
$$= 9.999 \text{ V}$$

Answer: The output voltage will be 10 volts.

To be an effective regulator the correct pulse width ratio is supplied to keep the output voltage constant. This is done by sampling the output voltage, comparing it to a reference voltage, and adjusting the mark/space ratio if necessary. Figure 10.4 is a simplified diagram of how this may be achieved. The error amplifier compares a proportion of the output voltage with the voltage at point V_D. If they are different an output is produced which adjusts the mark/space ratio till they are again equal.

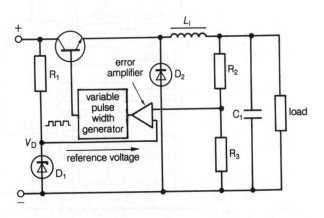

Fig. 10.4 *Basic action of an error amplifier in a switch mode voltage regulator circuit*

10.3 A practical switch mode regulator

Figure 10.5 illustrates the application of a switch mode regulator. The circuit is for a 5-volt voltage regulator with an output current of 1 ampere. The IC marked LM2535 contains all the functions required to provide pulses to the controlling switching BJTs Q_1 and Q_2. The IC has an internal regulated power supply, variable oscillator, error amplifier and current limiting. The external components are required to set up the output voltage and provide the current limit sensing.

The output voltage is set by the resistors R_1 and R_2. The terminal marked V_R is an internally generated voltage reference of 5 volts. Resistors R_4 and R_5 form a voltage divider between V_R and GND and the junction between R_4 and R_5 is fed to the non-inverting input of the error amplifier NI. The inverting input to the error amplifier INV is supplied from the voltage divider R_1 and R_2 which is connected across the output. Resistors R_4 and R_5 are equal in value, so the voltage at NI must be 2.5 volts. The output voltage, $V_{(OUT)}$, may be derived from the equation:

$$V_{(OUT)} = V_{NI} \left(1 + \frac{R_1}{R_2} \right)$$

where $V_{(OUT)}$ is the output voltage,
V_{NI} is the voltage at the INV terminal of the error amplifier,

R_1 and R_2 form the voltage divider across the output.

As R_1 and R_2 are both equal to 5 kilohms, in the circuit in Figure 10.5 the output voltage must be:

$$V_{(OUT)} = 2.5 \left(1 + \frac{5000}{5000} \right)$$
$$= 5 \text{ V}$$

So the output voltage is 5 volts.

The resistor, R_6, connected to the R_T terminal, and capacitor, C_1, connected to the C_T terminal, control the frequency of the internal oscillator which provides the pulses to the switching BJTs. In this case the frequency of the output pulses is 20 kilohertz. Capacitor C_2 and resistor R_7 determine the initial reference setting of the mark/space ratio. Resistor R_3 is the current sensing resistor and the two current sensing terminals CL+ and CL– connect across it. When the voltage across CL+ and CL– reaches 200 millivolts the current is reduced by 75 per cent. The terminals CA, EA, and CB, EB are the collector and emitter terminals of two internal BJTs A and B. In this circuit these are paralleled to drive the external paralleled BJTs Q_1 and Q_2.

The student is not required to completely understand the operation of the circuit and it is included here for information only. However the student should be able to recognise that this is a switch mode voltage regulator circuit.

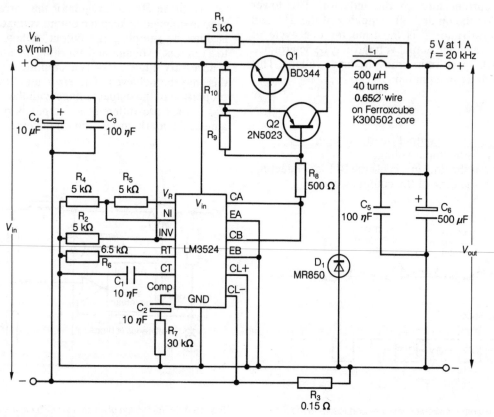

Fig. 10.5   *A practical switch mode step-down voltage regulator*

10.4 Step-up switch mode voltage regulator

A step-down switch mode regulator can deliver an output voltage no higher than the input. As the maximum amplitude of the 'on' pulses fed to the switching BJT is the input voltage, it must follow that the output at all times is less than the input. This, of course, is always the case for linear voltage regulators.

The simplified circuit in Figure 10.6 is of a *step-up* switch mode voltage regulator. Let us assume C_1 is charged as shown in the diagram and Q_1 is switched 'on'. Current flows as shown in the dashed lines—L_1 is storing energy in its magnetic field and C_1 is supplying current to the load. Point V_X is negative with respect to diode D_1, as when Q_1 is saturated the voltage drop across it is very small, so D_1 is reverse biased. Now when Q_1 is switched 'off' the fall in current through L_1 produces an emf of self induction which is in series with the supply voltage. Point V_X becomes positive, forward biases D_1, and current flows as shown by the solid lines both through the load and recharging C_1. The 'on' and 'off' times are such that the increasing current through L_1 when Q_1 is conducting does not reach a steady state value, and when Q_1 is not conducting the emf of self induction does not drop to too low a value. This produces an output voltage across the load which is the input voltage *plus* the average value of the emf of self induction.

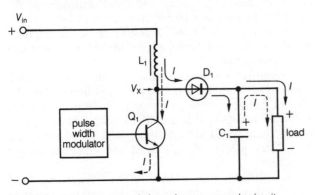

Fig. 10.6   *Basic step-up switch mode power supply circuit*

10.5 An inverting switch mode voltage regulator

A simplified circuit diagram of an *inverting switch mode voltage regulator* can be seen in Figure 10.7. Assuming C_1 is charged as shown in the diagram and Q_1 is conducting, current then flows as shown by the dashed lines, through L_1, storing energy in its magnetic field. At the same time C_1 delivers current to the load. Point V_X is positive and so diode D_1 is reverse biased. When Q_1 is switched 'off' and ceases to conduct, the emf of self induction produced in L_1 makes point V_X negative and causes current to flow through the load as well as recharging C_1, as shown by the solid lines.

It can be seen that the direction of current flow through the load is such that the common (lower) line becomes positive and the upper line negative. This is the reverse polarity of the input where the common line is negative. This type of regulator has an application where within a circuit there is a need to have negative potential with respect to common or earth.

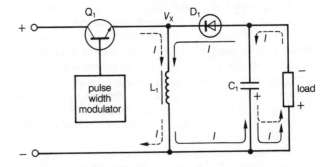

Fig. 10.7   *Basic inverting switch mode power supply circuit*

Unit 10 **SUMMARY**

- Linear power regulators usually operate at low efficiencies.
- A switch mode power supply operates on the principle of switching on and off a dc voltage.
- The switching, in a switch mode voltage regulator, is accomplished by cutting off or saturating a BJT.
- A switch mode voltage regulator samples the output voltage and compares it with a reference. The difference between the two is reduced to zero by adjusting the mark/space ratio of the switching action.

- A step-down switch mode voltage regulator produces an output voltage less than the input.
- A step-up switch mode voltage regulator produces an output voltage which is higher than the input.
- An inverting switch mode voltage regulator produces an output which has a reversed polarity, with respect to a common point, to an input.

Unit 11

Amplifiers 1—basic concepts

11.1 The nature of an amplifier

An amplifier is a device which raises the level of an input signal. By 'signal', we mean any constant or varying quantity which represents intelligent information. It is usually thought that an implifier must be electronic but in fact many other effects can produce amplifiers. These could be mechanical, pneumatic or hydraulic. One of the first public address amplifiers was operated by compressed air and was demonstrated by Sir Charles A. Parsons, the inventor of the steam turbine, in 1902. These compressed air amplifiers were very popular in Europe up to World War I and entertained people in concert halls, skating rinks and restaurants. However as this book deals with electronics we will confine the discussion to electronic types.

An amplifier is a *controlling device* in which the input signal controls an output. The power for the output is separate from the input so a small input signal may control the output in such a way that the output becomes an enlarged version of the input. This means that a lever, a megaphone, a magnifying glass or a transformer—even though they may all enlarge something—are not amplifiers, since they are not controllers, and no more power (in fact usually less) can come from them than has been put into them.

Although for many years all amplifiers used electron tubes, the basis of modern amplifiers is the transistor—either the BJT type discussed in Unit 6, or other types. The base current in a BJT can control the collector-emitter current so a small variation in base current can produce a larger, and similar, variation in collector-emitter current. The power supply for the collector is separate to the signal at the base and so the base current variation is *amplified* in the larger but similar collector current variation.

11.2 Amplifier symbols and functions

The general symbol for an amplifier is shown in Figure 11.1. This symbol does not signify by what internal process amplification occurs—simply that it does occur. The input enters the symbol on a base of the triangle and leaves, in an amplified version, at the opposite apex.

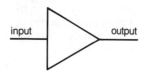

Fig. 11.1 *The simplified symbol for an amplifier*

The connection of an amplifier to an output load is shown in Figure 11.2. Here the symbol is connected to a signal source, represented by the sine wave in a circle, through the amplifier to an output load, represented by a resistor. As this is a very simplified diagram, no power supply for the amplifier symbol is shown. However it must be understood that an amplifier cannot function without a dc power supply, as it is the modification of the current drawn from the power supply that is the essence of amplification.

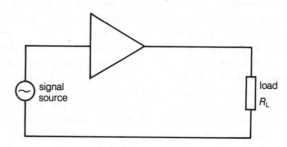

Fig. 11.2 *A simplified amplifier circuit, including input and output*

In addition to producing an output which is an enlarged version of the input, an amplifier must also be able to match itself between the input and output. This is accomplished by making the input resistance of the amplifier many times larger than the impedance of the signal source, and the output resistance of the amplifier equal to or less than the impedance of the load, depending on its application.

Amplifiers are broadly separated into two categories: small signal amplifiers and power amplifiers. While the division between these two is rather indeterminate, it is generally accepted that amplifiers with power outputs of less than 1 watt are classified *small signal amplifiers*. Amplifiers greater than 1 watt are classified *power amplifiers*.

Amplifiers may have many functions and classifications depending on their application. They may be used to amplify current changes (*current amplifiers*). They may be used to amplify voltage changes (*voltage amplifiers*). They may simply be used for impedance matching between a load and source of widely differing impedances. Power amplifiers could also be used for increasing power (for instance, to operate some mechanical device).

When an input causes an output to be similar to—but greater than—the input, we say a *signal gain* occurs. Remember this does not mean we are getting something for nothing, but that the input is controlling the output in a similar manner to itself. The gain of an amplifier is an indication of how much the input can control the output, so in the abovementioned amplifier classifications we may have *voltage gain, current gain* or *power gain*.

The mark of quality of an amplifier is its ability to amplify a signal such that the output is an exact enlarged version of the input. No amplifier can do this but some come very close—with output figures within 99.99 per cent of the input. We call this *percentage distortion* and where a distortion of 10 per cent may be adequate for some applications, others (especially high fidelity sound reproduction) may demand distortion levels of less than 0.01 per cent. (The term 'high fidelity', incidentally, means high faithfulness to the original sound.)

11.3 Amplifier construction

Electronic amplifiers may have many differing methods of construction. Older amplifiers may use electron tubes (valves), as previously mentioned, but modern types use almost exclusively solid-state electronic devices. These can be built up from few or many separate transistors. These circuits are usually termed *discrete component* circuits, as each part of the circuit is assembled from separate components. The circuit of a discrete component amplifier is shown in Figure 11.3—it is not expected that the student should understand this circuit, just recognise its type.

Complete amplifiers may be made up from ICs (integrated circuits) which may internally contain many individual components including transistors, resistors and capacitors. Figure 11.4 shows the circuit of a small IC amplifier. Note that the only external components are *passive devices*, i.e. resistors and capacitors, and all the *active devices*, e.g. transistors, are *internal* to the IC. The IC amplifier in the circuit has numbers about the triangle symbol, representing the numbered pins on the IC. This amplifier has an output of just over 1 watt and has a gain of 20. The gain may be increased, if required, by connecting an external resistor and capacitor between pins 2 and 6, otherwise they are left unconnected. This amplifier, when delivering an output of 500 milliwatts, has a distortion of 0.2 per cent. Once more it is emphasised that the student need not understand the working of the circuit, but just recognise its use.

Figure 11.5 shows the circuit of a hybrid amplifier which uses both an IC and separate BJTs. This amplifier will produce an output of 30 watts into a 4 ohm load. While it is possible to obtain ICs capable of high power outputs in themselves, many designers prefer to use a

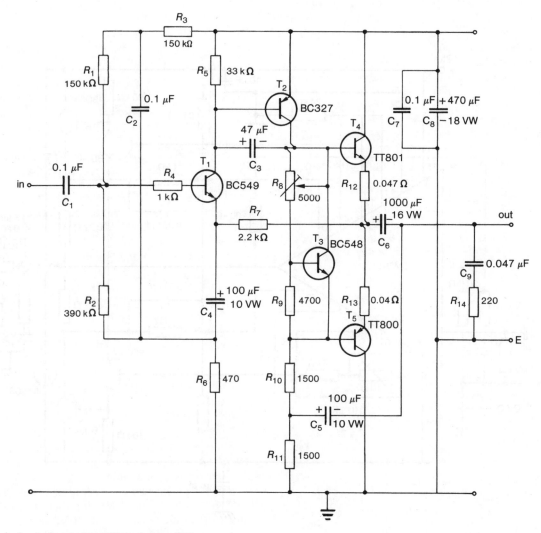

Fig. 11.3 *A circuit of a small amplifier using five BJTs*

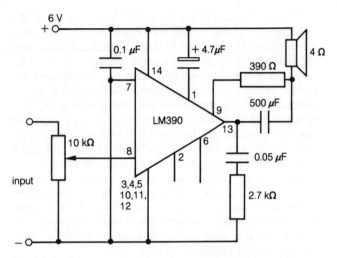

Fig. 11.4 *Circuit of a small amplifier using only a single IC*

low power IC to drive BJTs to obtain the power outputs required. Note the combination of PNP and NPN BJTs in the output. Do not be concerned with circuit operation but recognise that this amplifier circuit contains both an IC and four separate BJTs.

11.4 The voltage gain of an amplifier

Mention of the term 'gain' was made in Section 11.3. This is the control of an output by an input, and the gain figure of most importance is usually the voltage gain. In other words if a small varying voltage is applied to the input, a higher varying voltage of similar form is obtainable from the output.

It is the ratio of the output to the input that constitutes the numerical figure of gain. This ratio is given the symbol 'A' which is suffixed with the type of gain. As we are usually concerned with voltage gain the symbol for this would be A_V. The equation for voltage gain is as follows:

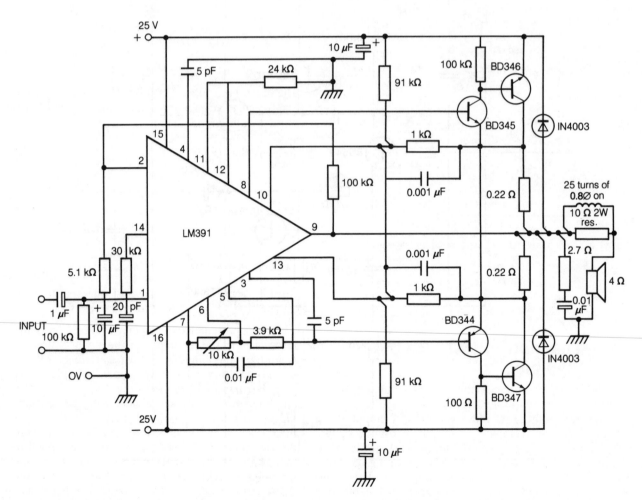

Fig. 11.5 *The circuit of a medium power amplifier using an IC as well as four BJTs, sometimes called a hybrid circuit*

$$A_V = \frac{v_o}{v_i} \qquad \textbf{(11.1)}$$

where A_v is voltage gain,
v_o is output voltage, and
v_i is input voltage.

Example 11.1

An amplifier, as represented in Figure 11.6, has an input voltage of 100 millivolts applied to the input terminals. At the output terminals the amplified signal is measured at 1 volt. What is the gain of the amplifier?

$v_i = 100\,\text{mV}$,
$v_o = 1\,\text{V}$,
$A_v = ?$

$$A_v = \frac{v_o}{v_i}$$
$$= \frac{1}{0.1}$$
$$= 10$$

The voltage gain of the amplifier is 10.

The voltage gain of an amplifier may vary greatly and can range from extremely high values, say 20,000, to less than 1. It may be argued that anything less than 1 could not possibly be a 'gain' but as it is a ratio of voltage output to voltage input the term remains. Some special amplifiers are made deliberately with a voltage gain of less than one, but they may have a considerable current gain.

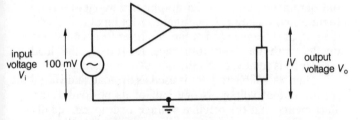

Fig. 11.6 *A diagram illustrating voltage gain in an amplifier*

11.5 The input resistance of an amplifier

In Section 11.2 mention was made of the input and output resistance of an amplifier and how an amplifier should be matched both to its input source and output load. For this reason the input and output resistance of the amplifier should be known. The input resistance is often described as the resistance seen when 'looking into' the input terminals of the amplifier (see Fig. 11.7). This input resistance may vary greatly depending on the type of amplifier circuit.

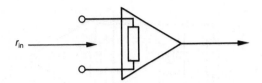

Fig. 11.7 *Representation of the input resistance of an amplifier*

The input resistance may be calculated from Ohm's Law in the following form:

$$r_i = \frac{v_i}{i_i} \qquad \textbf{(11.2)}$$

where r_i is the input resistance of the amplifier,
v_i is the input signal voltage, and
i_i is the input signal current.

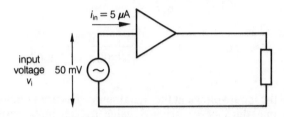

Fig. 11.8 *The input resistance of an amplifier could be determined from input voltage and input current*

Example 11.2

The diagram in Figure 11.8 represents the input voltage and current to an amplifier. Determine the input resistance of the amplifier.

$v_i = 50\,\text{mV}$,
$i_i = 5\,\mu\text{A}$, and
$r_i = ?$

$$r_i = \frac{v_i}{i_i}$$
$$= \frac{50 \times 10^{-3}}{5 \times 10^{-6}}$$
$$= 10\,\text{k}\Omega$$

Answer: The input resistance of the amplifier is 10 kilohms.

It is important that the input resistance of an amplifier is *matched* to the resistance of the signal source. If this is not done the amplifier may *excessively load* the signal source resulting in a lower input voltage, v_i, at the amplifier terminals. In effect the input resistance of an amplifier forms a voltage divider with the resistance of the signal source as shown in Figure 11.9. Using these figures let us see what voltage would be presented to the amplifier. The amplifier input voltage, v_i, will be the proportion of the signal voltage, v_s, as determined by the voltage divider action of the internal resistance of the signal source, r_s, and the amplifier input resistance, r_i, as follows:

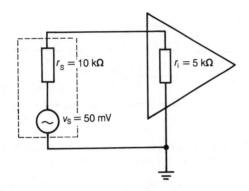

Fig. 11.9 *An input resistance lower than source resistance will excessively load the input signal*

$$v_i = v_s \left(\frac{r_i}{r_s + r_i} \right)$$

$$= 50 \times 10^{-3} \times \left(\frac{5000}{10000 + 5000} \right)$$

$$= 16.67 \text{ mV}$$

So the input voltage at the signal source has been reduced to one third of its original value by the input resistance of the amplifier itself. For this reason the input resistance of the amplifier should be at least ten times the resistance of the signal source. If it is at least this figure the voltage at the amplifier input will be reduced by a minimum of 9 per cent.

Now let us consider what effect an amplifier with an input resistance of 1 megohm will have on the voltage at its terminals under the same conditions of signal source as in Figure 11.9. Using the same voltage divider equation as above,

$$v_{in} = v_s \left(\frac{r_i}{r_s + r_i} \right)$$

$$= 50 \times 10^{-3} \times \left(\frac{1 \times 10^6}{10 \times 10^3 + 1 \times 10^6} \right)$$

$$= 49.5 \times 10^{-3} \text{ V}$$

So the reduction in input voltage is insignificant, being a difference of only 1 per cent. The amplifier can be said to be *matched* to the input source.

11.6 The output resistance of an amplifier

The *output resistance* of an amplifier could be described as the resistance seen looking 'back into' the amplifier from the output side. This is shown in the diagram in Figure 11.10.

Once more, as for the input resistance, we may calculate the output resistance from the Ohm's Law equation as follows:

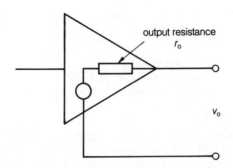

Fig. 11.10 *Representation of the output resistance of an amplifier*

$$r_o = \frac{\Delta v_o}{\Delta i_o} \qquad (11.3)$$

where r_o is output resistance,
Δv_o is change in output voltage, and
Δi_o is the change in output current.

The Δ symbol is used in the output resistance equation because the change in load resistance (which must accompany change in output current) will affect the amplifier gain. In general, the lower the output resistance, within limits, the better the performance of the amplifier. The load resistance also has a bearing on the amplifier gain—and once more, for voltage amplifiers, if the output resistance of the amplifier is much less than the load resistance, the output voltage will not be reduced. It must be noted here that in dealing with *power amplifiers* to get maximum output power, the load resistance must equal the output resistance of the amplifier.

It is possible, by experimental methods, to determine the output resistance of an amplifier without resorting to the above equation. Once again, as for input resistance, the output resistance (r_o) of the amplifier forms a voltage divider with the load resistance R_L. Firstly the load resistance is removed and the output voltage measured. Then a variable resistor is used as the load and varied till the output voltage is one half of its no-load value. This means that the original voltage is dropped equally across both sections of the voltage divider. In this case then, the two resistances are equal and so the amplifier output resistance is equal to the resistance of the variable resistor. This is shown in diagrammatic form in Figure 11.11. In (a) the output voltage is v_o and in (b) the output voltage is $v_o/2$, so r_o must equal R_L, which can be measured. If a decade box is used for R_L, the resistance may be read directly from the dials.

11.7 The equivalent circuit of an amplifier

No matter what form an amplifier may take (whether a single IC, a couple of transistors or a complex circuit containing many transistors and/or ICs), it may be represented by the simple triangle symbol. This symbol means that for a small input a usually much larger output

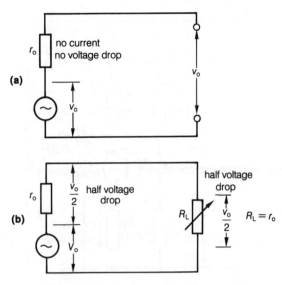

Fig. 11.11 *Illustration of experimental measurement of amplifier output resistance, as explained in the text*

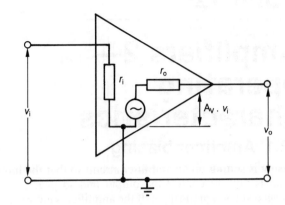

Fig. 11.12 *The equivalent circuit of an amplifier*

may be obtained; also, there is an input resistance and an output resistance. Remember that the input *controls* the output, and the output is provided by an external supply (usually not shown in the simple schematic symbol).

This means that we could consider the output to be a separate voltage (or current) generator, simply controlled by the input. This is the form usually expressed when we consider the *amplifier equivalent circuit*. Consider the equivalent circuit shown in Figure 11.12. Here we can see the input voltage, v_i, applied to the input resistance, r_i. This causes the output 'generator' to produce a voltage A_v times the input voltage, v_i, because A_v is the voltage gain of the amplifier. The output 'generator', through its own internal resistance, r_o, supplies the output voltage, v_o, that is applied to the load. As we have seen before, if the load resistance is many times greater than the output resistance the voltage drop across r_o is negligible.

Unit 11 **SUMMARY**

- An amplifier is any device in which an input can control a greater output.
- A simple equilateral triangle is the symbol used to represent any type of amplifier.
- An input to an amplifier is called the signal and the output is connected to a load.
- Amplifiers can be small signal amplifiers or power amplifiers.
- Three uses for amplifiers are: voltage amplifiers, current amplifiers and power amplifiers.
- The gain of an amplifier is the increase in output over the input. The most used gain is the voltage gain, A_v.

- Amplifiers may be made up from several single transistors, single ICs or a combination of both.
- The input resistance of an amplifier must be made much greater than the resistance of the signal source.
- The output resistance of a voltage amplifier must be kept many times less than the resistance of the load.
- The equivalent circuit of an amplifier shows the amplifier with input and output resistance and a controlled output generator.

Unit 12

Amplifiers 2— operating characteristics

12.1 Amplifier biasing

Biasing is setting up an amplifier circuit so that the input signal is able to control the output and so produce an enlarged version of the input. If the amplifier is incorrectly biased, the process of amplification will not take place or, if it does the output will be greatly distorted. Biasing may be as simple as the setting of the correct voltage supply, or as complicated as having to adjust most of the components of the amplifier circuit. If we are using a ready built amplifier, the only setting up required is to apply the correct supply voltage. If we are assembling an amplifier circuit from discrete components, the correct components must be selected before the amplifier will operate properly. If the correct components are used, the amplifier will be correctly biased by the supply voltage.

In Figure 12.1 we see a similar circuit to that shown in Figure 6.10. In this case the BJT has a 20 volt supply to the collector through resistor R_C. A 20 millivolt ac signal voltage is applied to the input but no collector current flows. This is the start of a very simple amplifier circuit using a single BJT. Before the BJT can act as an amplifier, however, we have to set up a bias current in the base circuit so that a steady or quiescent current will flow in the collector circuit.

Now in Figure 12.2 a voltage divider, R_1 and R_2, is connected across the supply (V_{CC}), and the base of the BJT is connected to the junction. This will allow base current to flow which, by transistor action, will permit collector current to flow. The collector current is such that the voltage between the output terminal and the common (earth or ground) is about one half the voltage of the supply.

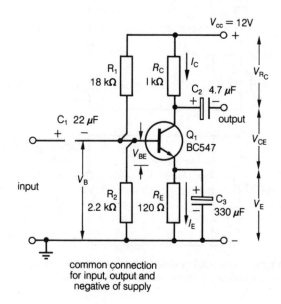

Fig. 12.2 *Simple single stage amplifier showing voltage points and current paths*

Let us briefly examine the conditions in the circuit of Figure 12.2 by calculating the currents and voltages. First find the voltage V_B. This is done by using the voltage divider equation,

$$V_B = V_{CC} \times \frac{R_2}{R_1 + R_2}$$

$$= 12 \times \frac{2.2 \times 10^3}{18 \times 10^3 + 2.2 \times 10^3}$$

$$= 1.307 \text{ V}$$

Now the base-emitter voltage drop is taken as 0.6 since the BC547 is a silicon BJT, so the difference between the base voltage and the base-emitter voltage drop is the emitter voltage drop, V_E.

$$V_E = V_B - V_{BE}$$

$$= 1.307 - 0.6$$

$$= 0.707 \text{ V}$$

By Ohm's Law we can now find the emitter current I_E.

$$I_E = \frac{V_E}{R_E}$$

$$= \frac{0.707}{120}$$

$$= 5.89 \times 10^{-3} \text{ A}$$

Now as emitter current is approximately equal to collector current (since base current is *very* small) we can say,

$$I_E = I_B + I_C$$

$$\simeq I_C$$

$$\simeq 5.89 \times 10^{-3} \text{ A}$$

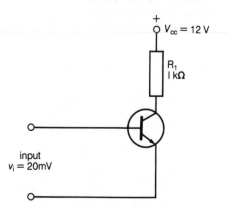

Fig. 12.1 *Without biasing there can be no amplification*

The voltage drop across resistor R_C can also be found by Ohm's Law,

$$V_{R_C} = I_C R_C$$
$$= 5.89 \times 10^{-3} \times 1 \times 10^3$$
$$= 5.89 \text{ V}$$

Now by Kirchhoff's Voltage Law the voltage drop across the BJT, V_{CE}, is

$$V_{CE} = V_{CC} - V_{R_C} - V_E$$
$$= 12 - 5.89 - 0.707$$
$$= 5.4 \text{ V}$$

and this is about one half of V_{CC} which is 12 volts.

Now when a steady (or quiescent) collector current is flowing, a *change* in the base current will cause a *change* in the collector current. Referring to Figure 12.3 it can be seen that the incoming ac signal can cause the steady base quiescent current to vary instantaneously above and below the quiescent value. This causes the collector current also to vary in the same manner. It can be seen that unless the bias conditions are correctly set, this may not be able to occur.

12.2 Signal clipping

It was mentioned in Section 12.1 that incorrect biasing may produce distortion in the output of an amplifier. Consider Figure 12.4, where the base current has been set too low. As the signal is applied, it causes the base current to drop to zero on the negative half of the incoming signal. This will cause the BJT to be cut off and, during this part of the output cycle, there can be no collector current. This results in a *clipping* of the output, and the output waveform does not follow the shape of the input signal and is thus distorted.

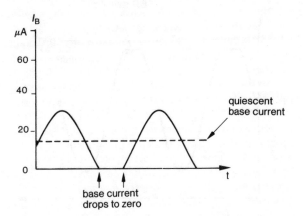

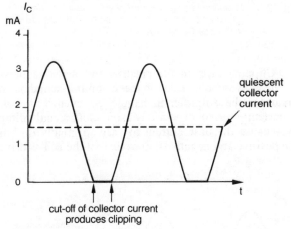

Fig. 12.4 *Cut-off in collector current is produced by having the quiescent bias current too low. This clips the wave on the negative half-cycle*

In Figure 12.5 we can also see clipping as a result of the base bias current being set too high. In this case, the higher base current causes a higher collector current to flow. Now the variation of the quiescent base current by the incoming signal causes the collector current to become so high, on the positive half of the signal waveform, that it drives the BJT into saturation. When the BJT is saturated it cannot pass any more current and so the output wave is once more clipped—but this time on the positive half cycle—again resulting in severe distortion.

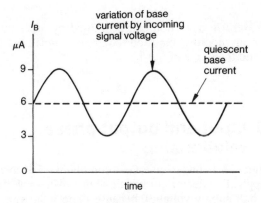

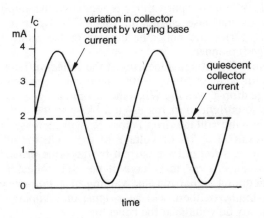

Fig. 12.3 *Input and output current variations in a BJT circuit produced by an incoming signal voltage, when correctly biased*

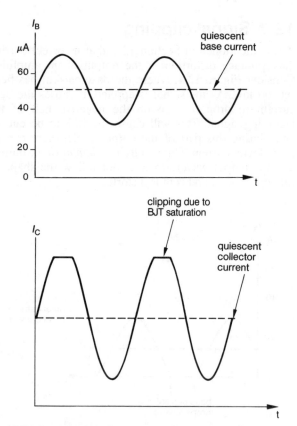

Fig. 12.5 *Saturation in collector current is produced by having the quiescent bias current set too high. This clips the wave on the positive half-cycle*

Clipping of both the positive and negative halves of the wave could occur if the incoming signal is too great for the amplifier to handle. So even if the bias is correctly set, an excessively large input signal voltage may cause the base current to vary so greatly on both the positive and negative half cycle that the BJT is driven

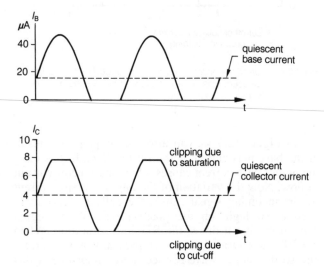

Fig. 12.6 *Clipping on both positive and negative peaks of the wave due to an amplifier being overdriven*

into both cut off *and* saturation. The result of this can be seen in the output waveform in Figure 12.6, which shows very severe distortion if it is assumed that the incoming signal was a sine wave.

When an amplifier is operating so that it is not clipping, we say that it is operating in a *linear* mode. If it is operated with incorrect bias, or *overdriven* by too large an input signal, we say that it is in non-linear operation. Figure 12.7 shows linear and non-linear operation using the simple amplifier symbol.

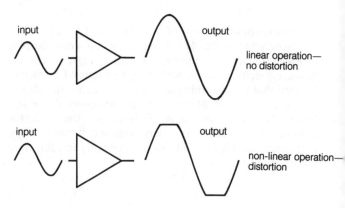

Fig. 12.7 *An overdriven, or badly biased, amplifier operates in a non-linear fashion*

Waveform distortion can be caused by other things apart from incorrect bias and too great a signal, but that is beyond the scope of this book.

12.3 Input and output phase relationship

In Figure 12.2 (under quiescent conditions—i.e. correct bias applied, collector current flowing, collector voltage about half supply voltage), because there is no varying voltage, there is no output, since capacitor C_2 blocks any dc. When an input voltage is applied to the amplifier it causes the base current to vary. This, by transistor action, varies the collector current in the same (but enlarged) manner.

Now the voltage appearing at the collector (between collector and negative) is the supply voltage *less* the voltage drop across R_C. When the incoming signal causes the base current to increase—and likewise the collector current—the voltage drop across R_C will increase. This will result in a *fall* of voltage at the collector of the BJT. Also, a *fall* in base current by the same procedure will produce a *rise* in voltage at the BJT collector. As the incoming signal may be varying at a frequency of say 1 kilohertz, there will be a ripple of 1 kilohertz on the steady dc voltage at the collector.

From the above it will be apparent that the *phase* of this voltage is *opposite* to the input voltage, or we can say there is a phase relationship of 180^0. As this is a varying voltage it will not be blocked by the capacitor, C_2, and will appear at the output terminal as a larger voltage than the input (i.e. it will be amplified) and of opposite phase. Another term for this is to say the output has been *inverted*.

If the output of a simple amplifier, such as that in Figure 12.2, was applied to the input of another simple amplifier (or as it is usually called an amplifier *stage*) the output would be inverted once more and so appear at the output of the second stage in phase with the original or *non-inverted* input. Figure 12.8 shows the input and output relationship of both inverting and non-inverting amplifiers.

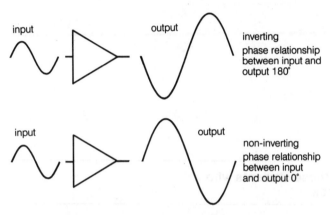

input output
inverting
phase relationship
between input and
output 180°

input output
non-inverting
phase relationship
between input
and output 0°

Fig. 12.8 *Output wave of inverting and non-inverting amplifiers in relation to their input waves*

12.4 Amplifier frequency response

The term *frequency response* refers to the ability of an amplifier to equally amplify a particular range of frequencies. Amplifiers are made for very many purposes and depending on their application may have very widely differing frequency responses. Audio (sound) amplifiers usually have a frequency response from just below, to somewhat above, the limits of human hearing—say from 30 hertz to 20 kilohertz.

Amplifiers in CROs may have a frequency response from dc to, say, 10 megahertz, a very wide bandwidth. Others may have a deliberately narrow frequency response to operate in a narrow band, for instance telephone circuits could be from 100 hertz to 4 kilohertz.

Amplifiers with very wide bandwidths are not common and usually there is a drop off in output at either end of their operating frequencies. This is often essential for their correct performance. Over their designed bandwidth the output is usually flat with a *roll-off* at each end and not an abrupt change. This can be seen in Figure 12.9. The bandwidth of an amplifier is taken as the range of frequencies between the two points at each end of the band which are 0.707 of the normal

output. These two points are often called the *break frequency* points. The lower break frequency point is marked as f_1 and the upper as f_2. The bandwidth of the amplifier is the range of frequencies between these two points. This may be expressed mathematically as,

$$BW = f_2 - f_1 \qquad (12.1)$$

where BW is the amplifier bandwidth,
f_2 is upper break frequency, and
f_1 is lower break frequency.

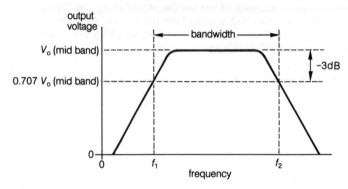

Fig. 12.9 *Curve of amplifier frequency response showing nominal bandwidth*

Example 12.1

An amplifier has a lower break frequency of 400 hertz and an upper break frequency of 3.5 kilohertz. What is its bandwidth?

f_1 = 400 Hz
f_2 = 3.5 kHz
BW = ?

$$
\begin{aligned}
BW &= f_2 - f_1 \\
&= 3500 - 400 \\
&= 3100 \text{ Hz}
\end{aligned}
$$

Answer: The bandwidth of the amplifier is 3.1 kilohertz.

The frequency break points are often called the *half-power points* or the *−3 dB points*. (The abbreviation 'dB' stands for *decibel* which is one tenth of a ratio called a *bel*; the derivation and the use of these terms need not concern us here, suffice to say they are a ratio between two levels of output voltage. The term bel was named after Alexander Graham Bell, the inventor of the telephone, and is based on the human perception of loudness of a sound in relation to the power of the sound source.)

Power is proportional to voltage squared, resistance being constant. If the output voltage, v_o, of an amplifier drops to 0.707 of its normal or midband level, the power output drops to 0.707^2 or 0.5 of the previous level. This is the 'half power' mentioned above and, in decibel terms, we say it is −3 dB below the previous level.

Unit 12 **SUMMARY**

- *Biasing* is the setting up of an amplifier so that an incoming signal will produce an undistorted output. This is provided by the dc supply voltage in a correctly set-up amplifier.
- If an amplifier is correctly biased it operates in a *linear* mode and does not clip the output waveform.
- If an amplifier is supplied with a signal voltage that is too high, the positive and negative peaks of the output are *clipped* and we say the amplifier is *overdriven*.
- In an *inverting* amplifier the output and input are 180° out of phase. In a *non-inverting* amplifier the output and input are in phase.

- The *frequency* response of an amplifier is its ability to equally amplify a particular range of frequencies.
- The *bandwidth* of an amplifier is the frequency range between two points at each end of the band which are the *half-power* points, or −3 *dB* of the normal mid-band output. These points are called the lower and upper *break frequency* points.

The following sections in this unit are only for those students who wish to study above the normal level of this book.

12.5 Quiescent stability

As can be seen in Figures 12.4 and 12.5 it is essential that the base bias current be correctly selected for the particular circuit under consideration and the value of the incoming signal.

It must be evident that to prevent clipping due to either BJT saturation or cut-off (or both), the quiescent conditions must be maintained. Any difference which may occur in a given circuit could be due to a change in h_{FE} of the BJT. This change may be due to a change in temperature or a change in the BJT (if it was replaced in the circuit).

If because of failure the BJT was replaced, even with one of the same type number, the dc current gain of the new BJT may not be exactly the same as the one replaced and this may lead to different quiescent values of currents. This could lead to distortion in the output unless the circuit components are altered to compensate. In later sections of this unit it will be shown how this effect can be overcome, but in a simple biased circuit it could pose a problem.

12.6 Effect of temperature on quiescent conditions

As has been shown in section 6.14, the current gain of a BJT can be greatly varied by variation in the junction temperature. If there is an increase in the junction temperature, the current gain (h_{FE}) of the BJT will increase, producing an increase in collector current for the same base current.

If in the circuit of Figure 12.10 there was an increase in temperature from, say, 30°C to 60°C, the collector current would increase to such a value that saturation would take place. This could be tabulated as follows:

Base current (I_B)   =   20 μA

At a junction temperature of 30°C,
collector current (I_C)   =   2 mA
collector voltage (V_C)   =   10 V

At a junction temperature of 60°C,
collector current (I_C)   =   3.5 mA
collector voltage (V_C)   =   2.5 V

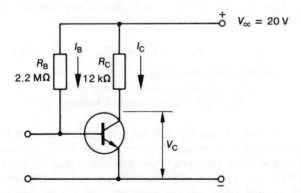

Fig. 12.10 *Simple base resistor bias circuit*

These conditions could completely upset the operation of the BJT and produce saturation distortion of the output voltage.

12.7 Simple base resistor bias

The simple base resistor bias circuit, as shown in Figure 12.10 is not stable, i.e. a change in current gain (h_{FE}) of the BJT in the circuit can completely alter the quiescent conditions. Stability may or may not be important, depending on the circuit application, but in many cases it must be considered.

The simple base resistor bias circuit is very simple to analyse and for this reason will be examined here in some detail before a more stable circuit is examined.

Let us examine the stability of the circuit in Figure 12.10. This entails an analysis of the circuit and the value of the collector voltage (V_C) with a change in h_{FE} of the silicon BJT from 100 to 180.

Using Kirchhoff's Voltage Law it may be stated that in Figure 12.10:

$$V_{CC} = I_B R_B + V_{BE}$$

This may be rewritten as:

$$I_B = \frac{V_{CC} - V_{BE}}{R_B} \qquad (12.2)$$

where I_B is base current,
V_{CC} is supply voltage,
V_{BE} is the base-emitter voltage drop,
R_B is the base resistor resistance.

For an h_{FE} of 100, using Equation 12.2:

$$I_B = \frac{V_{CC} - V_{BE}}{R_B}$$

$$= \frac{20 - 0.6}{2.2 \times 10^6}$$

$$= 8.8 \times 10^{-6} A$$

and, $I_C = h_{FE} I_B$

$$= 100 \times 8.8 \times 10^{-6}$$

$$= 0.88 \times 10^{-3} A$$

Now, $V_C = V_{CC} - I_C R_C$

$$= 20 - (0.88 \times 10^{-3} \times 12 \times 10^3)$$

$$= 9.4 \text{ V}$$

Answer: The collector voltage is 9.4 volts.

Now for an h_{FE} of 180:

$$I_B = \frac{V_{CC} - V_{BE}}{R_B}$$

$$= \frac{20 - 0.6}{2.2 \times 10^6}$$

$$= 8.8 \times 10^{-6} A \text{ (as before)}$$

and, $I_C = h_{FE} I_B$
$$= 180 \times 8.8 \times 10^{-6}$$
$$= 1.58 \times 10^{-3} A$$

Now, $V_C = V_{CC} - I_C R_C$
$$= 20 - (1.58 \times 10^{-3} \times 12 \times 10^3)$$
$$= 1.04 \text{ V}$$

Answer: The collector voltage is 1.04 volts.

It can be readily seen from these figures that the collector current has increased from 0.88 milliampere to 1.58 milliamperes and the collector voltage has decreased from 9.4 to 1.04 volts. These are entirely different quiescent conditions and in the second case the BJT would probably be saturated and could not function as an amplifier.

It can be seen that this circuit is not a stable circuit as the quiescent conditions are greatly influenced by a change in the current gain of the BJT.

12.8 Collector bias

The simple base resistor bias circuit can be slightly modified by reconnecting the base resistor as in Figure 12.11. In this connection the base current will be modified by the voltage drop across the resistor R_C. If the collector current tends to increase, the increased voltage drop ($I_C R_C$) reduces the base current and by transistor action automatically reduces the collector current.

This means that when the circuit is set up for a quiescent value of collector voltage, a change in current gain (h_{FE}) in the BJT, either due to temperature change or by replacing the BJT, will not have as great an effect as in the simple base resistor bias circuit. In other words, the *collector bias circuit* has a greater stability.

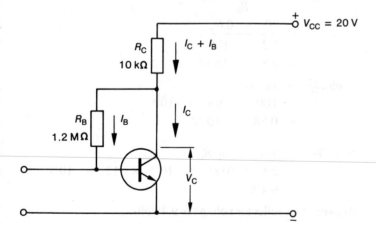

Fig. 12.11 *Collector bias circuit*

12.9 Voltage divider bias

A circuit with even better stability than the collector bias circuit is the *voltage divider bias* circuit. This circuit is also sometimes referred to as *emitter bias*, as the current

flowing in the emitter automatically adjusts the base bias current to keep the collector current constant.

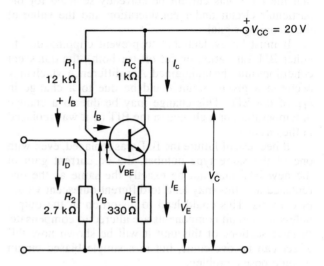

Fig. 12.12 *Voltage divider bias circuit*

Considering the circuit in Figure 12.12, it can be seen that the bias current is supplied from a voltage divider network, R_1 and R_2. Additionally, a resistor in the emitter circuit, R_E, assists in the biasing.

The circuit in Figure 12.12 will now be analysed in a non mathematical form, since the mathematics can be quite involved. Firstly the current through the voltage divider network (I_D) is made much larger than the base current. This is so because the base voltage (V_B) must be independent of small changes in base current. When operating from mains power, the voltage divider current is usually made 20 times the base circuit, but when operating from batteries is usually (for economy reasons) 10 times the base current.

The reason for the higher voltage divider current as compared to the base current may be explained by the fact that if I_D is 20 times the value of I_B and if, say, the base current changes by up to 10 per cent, there will be a change in base voltage, V_B, of about 0.5 per cent.

The base voltage is always equal to the emitter voltage plus the inbuilt base-emitter voltage (about 0.6 volt for silicon BJTs). This may be written as:

$$V_B = V_E + V_{BE} \tag{12.3}$$

where V_B is base voltage,
V_E is emitter resistor voltage drop,
V_{BE} is inbuilt base-emitter voltage.

As the current through the voltage divider may be considered to be constant, then the value of V_B will also be practically constant. Now, also, the value of V_{BE} is practically constant—thus from Equation 12.3, the value of V_E will be practically constant, and it can thus be said:

$$I_E = \frac{V_E}{R_E} \quad\quad (12.4)$$

where I_E is emitter current,

V_E is emitter resistor voltage drop,

R_E is the resistance in the emitter circuit,

and therefore the emitter current, I_E, will also be practically constant.

Thus changes in temperature will have little effect on the value of the emitter current and also, if the BJT is replaced with one having a differing current gain (h_{FE}), again there will be no appreciable change in emitter current. Now as collector current and emitter current are practically equal (since $I_E = I_C + I_B$ but I_B could be 1/200 of I_C), it can be said that there will be practically no change in the value of collector current.

Therefore the voltage divider bias circuit is a very stable circuit and, provided I_D is at least 10 times greater than I_B—and the current gain (h_{FE}) of the BJT is greater than 100—the quiescent conditions are independent of temperature and also variations in the current gain of different BJTs. It can also be said that the quiescent conditions are set by the supply voltage and resistor values, and not by the BJT.

Let us now consider what happens if, due to a temperature change, the gain of the BJT varies. The increased gain will tend to increase the collector current and therefore the emitter current. This will tend to increase the emitter voltage (V_E). Now, as the base voltage is constant, this will tend to reduce the value of base-emitter voltage (V_{BE}), which in turn will tend to reduce the current, which by transistor action will return the emitter current and collector current to their former value. In other words, quiescent stability is maintained.

In the design of the voltage divider bias circuit certain considerations are made. They are, first, the collector voltage and the collector current. It is usual to have the collector voltage about 0.5 of the supply voltage, so this fixes the supply voltage. Now the voltage across the emitter resistor is usually between 0.1 and 0.2 of the supply voltage.

Example 12.1

It is required to design a voltage divider bias circuit for a common emitter connected BJT. The collector current is to be about 10 milliamperes and the collector voltage about 10 volts. The BJT must have a design centre value of collector current of 10 milliamperes and an h_{FE} of about 150. It is a silicon type.

Determine the required collector and emitter resistor values and lower and upper voltage divider resistance values.

$$V_C = 10 \text{ volts}$$
$$I_C = 10 \text{ milliamperes}$$
$$\begin{aligned} V_{CC} &= 2V_C \\ &= 2 \times 10 \\ &= 20 \text{ V} \end{aligned}$$

$$\begin{aligned} V_E &= 0.15\ V_{CC} \\ &= 0.15 \times 20 \\ &= 3 \text{ V} \end{aligned}$$

$$\begin{aligned} \text{Now, } I_C\,R_C &= V_{CC} - V_C \\ &= 20 - 10 \\ &= 10 \text{ V} \end{aligned}$$

$$\begin{aligned} R_C &= \frac{V_{CC} - V_C}{I_C} \\ &= \frac{20 - 10}{10 \times 10^{-3}} \\ &= 1000\ \Omega \end{aligned}$$

Answer 1: The collector resistor will be 1 kilohm.

$$\text{Now, } I_E \simeq I_C$$
$$\begin{aligned} \text{so, } R_E &= \frac{V_E}{I_E} \\ &= \frac{3}{10 \times 10^{-3}} \\ &= 300\ \Omega \end{aligned}$$

Answer 2: The emitter resistor will be 330 ohms (nearest preferred E12 value).

Now I_D should be about 20 times I_B and I_B will be about $1/150\ I_C$, i.e.,

$$\begin{aligned} I_B &= \frac{I_C}{h_{FE}} \\ &= \frac{10 \times 10^{-3}}{150} \\ &= 6.7 \times 10^{-5}\text{A} \end{aligned}$$

$$\begin{aligned} \text{so, } I_D &= 20\ I_B \\ &= 20 \times 6.7 \times 10^{-5} \\ &= 1.34 \times 10^{-3}\text{A} \end{aligned}$$

$$\begin{aligned} \text{Now, } V_B &= V_E + V_{BE} \\ &= 3 + 0.6 \\ &= 3.6 \text{ V} \end{aligned}$$

$$\begin{aligned} \text{and, } R_2 &= \frac{V_B}{I_D} \\ &= \frac{3.6}{1.34 \times 10^{-3}} \\ &= 2686\ \Omega \end{aligned}$$

Answer 3: The lower voltage divider resistor will be 2.7 kilohms (nearest preferred E12 value).

Now the current in R_1 will be $I_D + I_B$, and the voltage drop across R_1 will be $V_{CC} - V_B$, so,

$$\begin{aligned} R_1 &= \frac{V_{CC} - V_B}{I_D + I_B} \\ &= \frac{20 - 3.6}{1.34 \times 10^{-3} + 6.7 \times 10^{-5}} \\ &= 11\ 656\ \Omega \end{aligned}$$

Answer 4: The upper voltage divider resistor will be 12 kilohms (nearest preferred E12 value).

The above calculations are approximate and are included for illustration purposes but are sufficiently accurate for normal design and are within the normal tolerance of component values. In fact, in the above calculation, the value of I_B in the voltage divider upper resistor current could be ignored.

Example 12.2

Determine the quiescent collector current and collector voltage for the circuit shown in Figure 12.13.

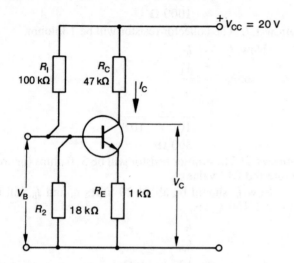

Fig. 12.13   *Voltage divider bias circuit for Example 12.2*

The BJT is a silicon type, which is normal.

$$V_B = V_{CC} \frac{R_2}{R_1 + R_2}$$

$$= 20 \times \frac{18 \times 10^3}{100 \times 10^3 + 18 \times 10^3}$$

$$= 3.05 \text{ V}$$

$$V_E = V_B - V_{BE}$$

$$= 3.05 - 0.6$$

$$= 2.45 \text{ V}$$

$$I_C = \frac{V_E}{R_E}$$

$$= \frac{2.45}{1000}$$

$$= 2.45 \times 10^{-3} \text{A}$$

Answer 1: The collector current is 2.45 milliamperes.

$$V_C = V_{CC} - I_C R_C$$

$$= 20 - (2.45 \times 10^{-3} \times 4.7 \times 10^3)$$

$$= 8.49 \text{ V}$$

Answer 2: The collector voltage is 8.5 volts.

Unit 13

Operational amplifiers 1— basic concepts

13.1 The basics of an operational amplifier

There is no such thing as a perfect amplifier but modern *operational amplifiers* come very close. These were developed originally for use in analog computers where they perform complicated mathematical operations, and this is how they were first given their name. Quite often the name is shortened to *op. amp.*

The ideal operational amplifier should have the following features:

1. an infinitely high input resistance;
2. an infinitely high voltage gain;
3. a zero output resistance.

Operational amplifiers are invariably made as integrated circuits, or ICs, and because of this no internal adjustments or modifications are possible. Additionally, the internal circuit is of little importance and the user relies on the manufacturer's data when designing circuits with IC op. amps. The internal circuit of most operational amplifiers consists of three sections with as many as 20 internal transistors to achieve the approach to the ideal

conditions stated above. For interest's sake only, the circuit of a typical IC op. amp. is shown in Figure 13.1.

The circuit of Figure 13.1 shows the positive and negative supply, an output terminal, two input terminals and two terminals marked 'offset null'. The last two are simply used with a variable external resistor to set up the output at zero volts when there is no input. The two input terminals are marked 'non inverting' and 'inverting'.

Perhaps the best way to represent the operational amplifier is to use the simplified symbol, and Figure 13.2 shows the simple triangle symbol with the two input terminals and the output. It is understood that the op. amp. must have a power supply and that offset null has been carried out.

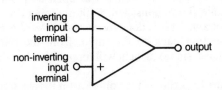

Fig. 13.2 *A simplified symbol of an operational amplifier*

The inverting input terminal will give an output with a phase difference of 180^0 to the input, and the non-inverting input will give an output that is in phase with the input. These are respectively marked – and + on the diagram in Figure 13.2

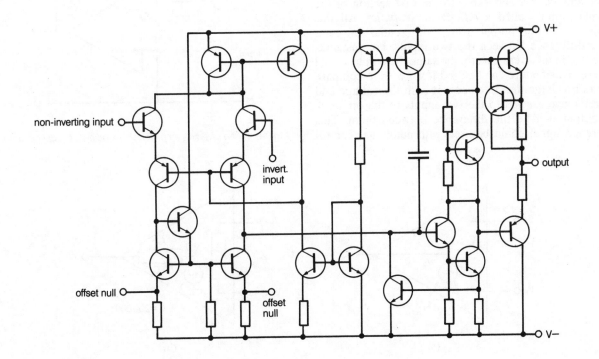

Fig. 13.1 *The internal circuit of a typical operational amplifier*

13.2 A practical operational amplifier

A typical operational amplifier is that designated μA741 by one manufacturer. Let us compare its practical characteristics with the ideal characteristics set out in Section 13.1.

1. Input resistance—2 megohms.
2. Open-loop voltage gain—200 000.
3. Output resistance—75 ohms.

While not 'ideal', these results are very impressive.

The maximum output voltage swing of this op. amp. is ± 10 volts, otherwise the output is driven into cut off and saturation, i.e. waveform clipping. This means that with a gain of 200,000 any input voltage greater than 50 microvolts will overdrive the amplifier. Because of this it is usual to reduce the open-loop gain to a manageable figure consistent with the input and desired output voltage. This is done by providing *feedback*, which means feeding back to the input, from the output, an out-of-phase signal which will oppose and reduce the input.

Figure 13.3 shows the comparison between the ideal and practical operational amplifier as far as input and output resistance is concerned.

Some operational amplifiers are made to operate from a single-polarity power supply, while others require a dual-polarity supply. Once again these are not normally shown in circuit diagrams, as knowledge of the type being used will indicate whether a single or dual-polarity supply is needed. (When following a wiring diagram, of course, the power supply details are included.)

The two inputs to an op. amp., shown in Figure 13.3 as V_A and V_B, are connected to the first section of the internal circuit called a *differential amplifier*. All this means is that the output from a differential amplifier is the difference between the two inputs. For instance, if the two inputs were exactly the same the output would be zero. This feature can be used if we wish to compare a certain voltage to a reference. If both the voltage and the reference are fed to the two inputs of the op. amp. the output will be the difference between them. This difference signal could be used with additional control

circuitry, for instance to change the voltage till it equalled the reference.

Other uses for op. amps include active filters (low pass, high pass, bandpass and notch), general purpose amplifiers (both inverting and non-inverting), oscillators, power supply voltage controllers, summing amplifiers and metering circuits.

13.3 Using operational amplifiers

It was mentioned in Section 13.2 that the operational amplifier has a very high *open loop gain*. The term 'open loop' is given when there is no feedback circuit (or loop) between input and output. In almost all operations a feedback loop is used so that the voltage gain may be controlled. When used as an inverting amplifier a resistor is connected between the output and inverting input terminal.

Additionally, another resistor is placed between the signal source, V_i, and the inverting input terminal. These two resistors form a voltage divider for the negative feedback voltage, and the reduced gain is the ratio of the feedback to the input resistance.

The two resistors controlling the voltage gain can be seen in the diagram in Figure 13.4. The feedback resistor is marked R_f and the input resistor is marked R_1. The inverting input terminal of the op. amp. in this

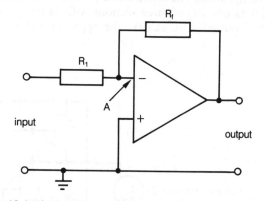

Fig. 13.4 *An inverting amplifier circuit using a typical op. amp.*

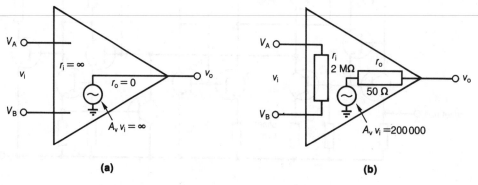

(a)   **(b)**

Fig. 13.3 *Comparison of (a) an ideal and (b) a practical typical operational amplifier*

connection is termed the *virtual earth* or *virtual ground* and is marked 'A' on the diagram. This is explained by always considering the op. amp. to have ideal characteristics. This means that as the input resistance is considered infinite, there is no input current—so it can be said $i_o = 0$ amperes. If there is no current between the inverting and non-inverting terminals the voltage at each terminal must be the same. Now if the non-inverting terminal is earthed, as in the diagram, the inverting terminal is also at earth (or ground) potential. Although the operational amplifier is considered to have infinite input resistance, the input resistance of the circuit is R_1.

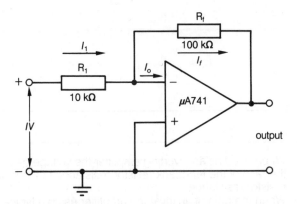

Fig. 13.5 *A practical inverting amplifier circuit using an operational amplifier*

In the circuit of Figure 13.5 the two resistors of Figure 13.4 have been assigned values. It can be seen that the circuit input resistance must be 10 kilohms. The voltage applied is 1 volt dc, so using capitals for the dc values, input current is

$$I_i = \frac{1}{10 \times 10^3}$$
$$= 100 \ \mu A$$

So 100 microamperes will flow into the circuit.

Now, as

$$I_o = 0 \ A$$
then $$I_f = I_i$$
$$= 100 \ \mu A$$

The output voltage,

$$V_o = -I_f R_f$$
so, $$V_o = -100 \times 10^{-6} \times 100 \times 10^3$$
$$= -10 \ V$$

[Note: I_f is flowing *into* the output terminal, so is considered negative.]

The output voltage is −10 volts.

Now the voltage gain, A_v, is equal to output voltage divided by input voltage (Equation 11.1), so

$$A_v = \frac{V_o}{V_i}$$
$$= \frac{-10}{1}$$
$$= -10$$

The voltage gain is −10. The minus is added to indicate the output voltage is inverted.

Now we have said that,

$$V_o = -I_f R_f$$
and $$V_i = I_i R_1$$
so $$A_v = \frac{-I_f R_f}{I_i R_1}$$
but as $$I_f = I_i$$
then $$A_v = -\frac{R_f}{R_1} \qquad (13.1)$$

where A_v is voltage gain,
R_f is feedback resistor resistance, and
R_1 is input resistor resistance.

The above derivation of the voltage gain equation for the op. amp. inverting amplifier circuit employed dc voltages at the input. This shows that the amplifier will amplify dc values. In addition, of course, the amplifier circuit will amplify ac signals. When the gain of the amplifier is reduced from its open loop condition the frequency response bandwidth of the amplifier increases. When on open loop, the typical op. amp. will amplify virtually only dc, as the upper break frequency is only about 10 hertz. At a voltage gain of 100 the upper break frequency is about 10 kilohertz, increasing to nearly 100 kilohertz when the voltage gain is reduced to 10.

If the feedback resistor is made the same value as the input resistance, it must be apparent from Equation 13.1, that the voltage gain of the amplifier circuit would be unity. This means that there is no increase in voltage between input and output, but sometimes this condition is desirable. At unity gain the circuit in Figure 13.5 would have a break frequency of about 1 megahertz.

Example 13.1

If, in the diagram in Figure 13.5, R_f has a resistance of 120 kilohms and R_1 a resistance of 8.2 kilohms, what is the voltage gain of the circuit?

$$R_1 = 8.2 \ k\Omega$$
$$R_f = 120 \ k\Omega$$
$$A_v = ?$$

$$A_v = -\frac{R_f}{R_1}$$
$$= -\frac{120 \times 10^3}{8.2 \times 10^3}$$
$$= -14.6$$

Answer: The voltage gain of the circuit is −14.6.

Example 13.2

If, in the circuit of 13.5, the resistance of R_f is increased to 1 megohm, what would be the voltage gain?

R_1 = $10 \text{ k}\Omega$

R_f = $1 \text{ M}\Omega$

A_v = ?

$$A_v = -\frac{R_f}{R_1}$$

$$= -\frac{1 \times 10^6}{10 \times 10^3}$$

$$= -100$$

Answer: The voltage gain of the circuit is -100.

Although the input resistance in the circuit of Figure 13.5 may be varied by changing the resistance value of R_1, the output resistance remains substantially the same as that of the op. amp., typically about 75 ohms.

Unit 13 **SUMMARY**

- Operational amplifiers come close to being the perfect amplifier.
- An operational amplifier has two inputs; an inverting input and a non-inverting input.
- Operational amplifiers are used for such purposes as: general purpose inverting and non-inverting amplifiers, differential amplifiers, summing amplifiers, comparators, active filters, oscillators, power supply regulators and metering circuits, in addition to many other uses.

- When used as an inverting amplifier the voltage gain is the ratio of the feedback resistor resistance to the input resistor resistance.
- When the gain of an inverting amplifier using a typical op. amp. is set at 10, the amplifier will have a bandwidth from dc to 100 kilohertz.

Unit 14

Operational amplifiers 2— using op. amps

14.1 Non-inverting amplifiers

The non-inverting amplifier circuit, using an operational amplifier, produces an output that is in phase with the input. In Unit 13 the only amplifier circuit shown was an inverting type. In the inverting amplifier the inverting terminal, usually marked with a minus (–) sign, is used as the input. The output is inverted for dc as well as phase reversal for an ac signal. This means that if the input was positive with respect to common (or earth), the output would be negative with respect to common. Also, if the input was negative with respect to common, the output would be positive. The non-inverting amplifier, as its name suggests, produces an output with the same polarity or phase as the input.

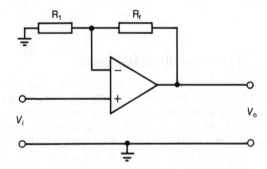

Fig. 14.1 *The basic circuit of a non-inverting amplifier*

In the diagram in Figure 14.1 it can be seen that the input is connected between the non-inverting input terminal (+) and common (earth). Feedback, which controls voltage gain, is provided through the voltage divider action of resistors R_f and R_1, the junction of which is applied to the inverting input terminal (–). This is the basic non-inverting amplifier circuit.

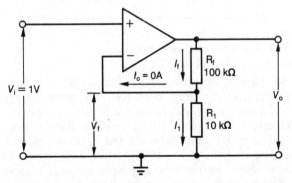

Fig. 14.2 *A non-inverting amplifier showing voltages and current paths*

In Figure 14.2 the amplifier symbol has been inverted so that the inverting input terminal is on the lower side. This gives a somewhat neater layout than the arrangement in Figure 14.1, as the two resistors R_f and R_1 connect across the output in a more recognisable voltage divider layout. Connected to the input is a 1 volt source which we will use, with Ohm's Law, to establish the voltage gain for this circuit.

Referring to Figure 14.2 we recall that if the voltage at two input terminals of an op. amp. is equal, there will be no output and no current will flow into these terminals because we consider the input resistance to be infinite. Now the output voltage, V_o, will adjust itself (with the voltage divider R_f and R_1) until the voltage V_f becomes equal to the input voltage V_i, which in our case is 1 volt. So we can say

$$V_f \ = \ V_i \ = \ 1 \text{ V}$$

Now, as resistor R_1 is across V_f, we can calculate the current in R_1:

$$\begin{aligned} I_1 \ &= \ \frac{V_i}{R_1} \\ &= \ \frac{1}{10 \ \times \ 10^3} \\ &= \ 100 \ \mu\text{A} \end{aligned}$$

The current through R_1 is 100 microamperes and as the input current is considered zero, we can say

$$I_f \ = \ I_1 \ = \ 100 \ \mu\text{A}$$

Now the voltage drop across the upper half of the voltage divider, R_f, is

$$\begin{aligned} V_{Rf} \ &= \ I_f R_f \\ &= \ 100 \ \times \ 10^{-6} \ \times \ 100 \ \times \ 10^3 \\ &= \ 10 \text{ V} \end{aligned}$$

and we can add this 10 volts to the voltage drop across R_1, to get the output voltage V_o:

$$\begin{aligned} V_o \ &= \ V_{Rf} \ + \ V_f \\ &= \ 10 \ + \ 1 \\ &= \ 11 \text{ V} \end{aligned}$$

So the output voltage is 11 volts. Now, from the voltage gain equation 11.1:

$$\begin{aligned} A_v \ &= \ \frac{V_o}{V_i} \\ &= \ \frac{11}{1} \\ &= \ 11 \end{aligned}$$

So the voltage gain of the non-inverting amplifier circuit in Figure 14.2 is 11, and because the figure is positive the output is in phase with the input.

It would be apparent that the voltage gain of a non-inverting amplifier is set by the ratio of the resistance of the two resistors R_f and R_1, so let us (for the general case) derive an equation for voltage gain using these values.

Once again referring to Figure 14.2, it can be seen that the voltage across R_1, V_f, is

$$V_f = V_o \left(\frac{R_1}{R_f + R_1} \right)$$

and

$$\frac{V_f}{V_o} = \frac{R_1}{R_f + R_1}$$

Inverting both sides,

$$\frac{V_o}{V_f} = \frac{R_f + R_1}{R_1}$$

and, as $V_f = V_i$

then

$$\frac{V_o}{V_i} = \frac{R_f + R_1}{R_1}$$

But, as

$$A_v = \frac{V_o}{V_i}$$

then also

$$A_v = \frac{R_f + R_1}{R_1}$$

$$= \frac{R_f}{R_1} + \frac{R_1}{R_1}$$

$$= \frac{R_f}{R_1} + 1$$

So it may be stated that the voltage gain of a non-inverting amplifier, using an op. amp., can be determined from the equation:

$$A_v = \frac{R_f}{R_1} + 1 \qquad \textbf{(14.1)}$$

where A_v is the voltage gain of a non-inverting amplifier,

R_f is the resistance of the feedback resistor, and

R_1 is the resistance of the voltage divider resistor.

Example 14.1

For the circuit in Figure 14.3 determine the voltage gain of the depicted amplifier.

R_f = 82 kΩ

R_1 = 6.8 kΩ

A_v = ?

$$A_v = \frac{R_f}{R_1} + 1$$

$$= \frac{82 \times 10^3}{6.8 \times 10^3} + 1$$

$$= 13.05$$

Answer: The gain of the amplifier is 13.

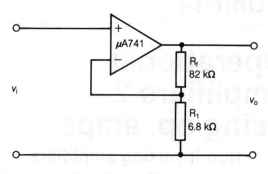

Fig. 14.3 *Circuit for Example 14.1*

As we consider the input resistance of an operational amplifier to be infinite, the input resistance of the non-inverting amplifier must also be infinite. In practice the input resistance of the non-inverting amplifier may be in the vicinity of 10 000 megohms, which can be considered as being infinite for all practical purposes. The non-inverting amplifier is sometimes called a *high input impedance amplifier.*

The output resistance of the non-inverting amplifier is very low and could be in the order of 0.005 ohm. This can be considered as being zero for all practical purposes.

14.2 The voltage follower

Consider the circuit in Figure 14.4. Here the input is applied to the non-inverting input terminal of an op. amp. and there is 100 per cent feedback applied from the output to the inverting terminal. This is, possibly, the simplest circuit which could be used. The gain of the circuit is 1, which (it may be argued) is not a gain—but when there is a control of an output by an input we say a gain occurs. No gain would imply no output.

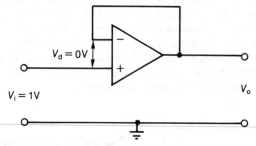

Fig. 14.4 *The voltage follower circuit*

We have seen in Section 14.1 that the two input terminals of an op. amp. must have the same voltage applied when a feedback connection is made. In the case of the circuit in Figure 14.4 we can see that, between the non-inverting input terminal and common, there is 1 volt and that the terminal is positive with respect to common. Now, for the inverting input terminal to have the same potential as the non-inverting one, the output voltage must rise to 1 volt when the system becomes stabilised. This then gives us the same voltage, and in the same phase, as the input. This is why the circuit is called a *voltage follower:* the output follows (almost exactly) the input.

The student may wonder at this stage just what use can be made of a circuit in which the output is exactly the same as the input. The value of the voltage follower circuit is that it has an extremely high input resistance (possibly limited only by the insulation between the terminals), and an extremely low output resistance which could be considered zero. Because of this it is used to link two circuits where one has a very high output resistance and the other a low input one. (If the low resistance input was connected to the very high resistance output it would load it so much that the output voltage would fall to an unusable level. If the voltage follower, however, is connected between the two circuits then the resistances are matched.) When used in this way the voltage follower is sometimes referred to as a *buffer amplifier*, as it acts as a 'buffer' between two incompatible circuits. This process is also termed impedance matching.

14.3 Three operational amplifier configurations compared

Table 14.1 compares the three circuits already discussed.

Table 14.1 *Comparison of amplifier circuits*

Type of amplifier	Input resistance	Output resistance	Voltage gain, A_v	Output polarity or phase
inverting	medium—depends on input resistor	low	$= \dfrac{R_f}{R_1}$	reversed or 180^0 out of phase
non-inverting	very high	very low	$= \dfrac{R_f}{R_1} + 1$	non-reversed or in phase
voltage follower	extremely high	extremely low	unity	non-reversed or in phase

14.4 Applications of op. amp. circuits

There are countless applications for the three circuits discussed in this unit and in Unit 13, so we will look at only typical uses. It is not suggested that the student should completely understand the circuit operations, just that they be noted.

The *voltage follower* can be used between low and high impedance circuits. In some circumstances a signal source will have a very high inherent resistance. An example of this could be a piezo-electric strain gauge cell. The distortion of the piezo-electric crystal produces a voltage across each face. This voltage is proportional to the degree and rate of distortion. The distortion could be produced by say, the action of a vibration in a part of a machine where the cell was mounted. Because the resistance between each face is extremely high, any low resistance connected across it would reduce the voltage to a very low figure.

If the output of the piezo cell is connected to the input of a voltage follower, the same voltage appears at the output of the voltage follower but with an extremely low output resistance which could be directly connected to a low impedance control circuit (possibly mounted some distance away and connected by long circuit conductors). The control circuit could sound an alarm if vibration became excessive, by responding to a certain maximum voltage level. Such a circuit appears in Figure 14.5.

The *non-inverting* amplifier could be used to amplify any small voltage source, be it dc voltage changes or ac. Like the voltage follower it has a very high input resistance and will not load a circuit to which it is connected. One example of this application is in amplifying the very small voltages produced by nerve impulses in the human body. This could be used to amplify the voltages produced by heart action and assist in driving an electro-cardiograph which can trace out the electrical impulses from a heartbeat to draw a graph of the heart action. This could show up any abnormality in the heart.

A circuit for this application appears in Figure 14.6. The LH0036 consists of three op. amps in the one package. One op. amp. uses pins 4 and 5 with pin 4 the inverting input and pin 5 the non-inverting input.

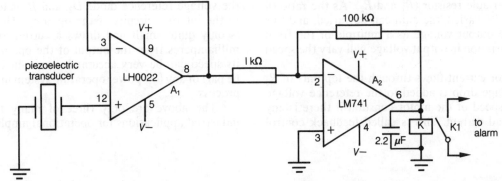

Fig. 14.5 *A voltage follower circuit, A_1, used to match the very high resistance of the piezoelectric transducer to the operating section of the circuit A_2*

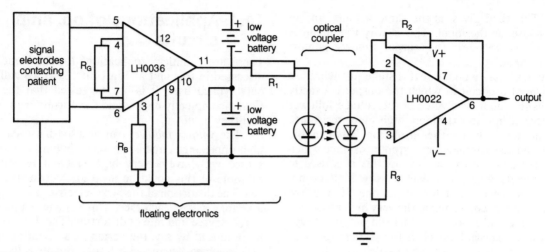

Fig. 14.6 *Non-inverting amplifiers used in medical electronics*

Likewise pin 7 is non-inverting and pin 6 inverting for the second op. amp. The outputs from the first two op. amps feed the two inputs of the third. Resistor R_G controls the gain and R_B the bias. Both these depend on where the electrodes are placed on the patient.

The electronic circuit connected to the patient is supplied from a centre-tapped low-voltage battery for safety. The connection to the recording circuit is made through an isolating opto coupler (these are examined in section 27.5), being further amplified by op. amp. LH0022. Both these ICs are termed instrumentation amplifiers and are op. amps designed for precision differential signal processing. Resistor R_G can vary the gain up to 1000 and the input resistance is about 300 megohms.

As the *inverting* amplifier has only a medium input resistance, which could drop to say 100 ohms at very high gain, it cannot be used to match a high source impedance. However, because of its inverting action, an increase in a dc voltage at the input will show as an increase in the negative output. This action could be used to provide a negative supply to operate a motor, and a circuit doing this is shown in Figure 14.7.

In Figure 14.7 the input to the amplifier is fixed (V_{ref}). This reference voltage is the voltage drop across diode D_2. The feedback and input resistor are both part of the 10 kilohm variable resistor (R_1 and R_2). As the ratio of R_2 to R_1 can be varied, the gain can be varied, and this will vary the output voltage as a multiple of the fixed input. The variation in output voltage will vary the speed of the motor.

The motor current flows through the top part of R_3 and this voltage drop is added to the reference voltage to keep the speed of the motor constant if there is any change in load current. This is called feedback control

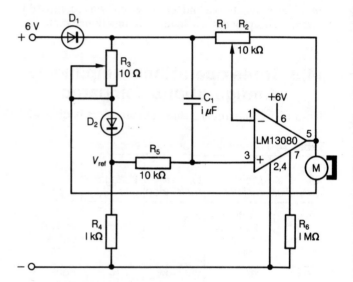

Fig. 14.7 *An inverting power operational amplifier used to vary and control the speed of a small motor*

and the amount of control can be preset by adjusting R_3.

C_1 and R_5 are simply filter components for the reference voltage. R_4 provides a forward bias current for the voltage reference diode D_2, and R_6 is to supply bias to the internal circuits of the op. amp. The motor here is only quite small and draws a current of about 250 milliamperes from the output of the op. amp. However its speed can be very accurately controlled and it could be part of any low power operating system in an industrial process.

The above are only three of many thousands of industrial applications for operational amplifiers.

Unit 14 **SUMMARY**

- An inverting amplifier produces an output of opposite polarity or phase reversal.
- A non-inverting amplifier produces an output with the same polarity, or is in-phase with the input.
- A non-inverting amplifier has a very high input resistance and a very low output resistance.
- The voltage follower has a gain of 1 but has an extremely high input resistance and an extremely low output resistance.

- The voltage follower is used as a buffer between high and low impedance circuits.
- There are many thousands of industrial uses for amplifier circuits using operational amplifiers.

Trigger devices for thyristors

15.1 The UJT (unijunction transistor)

The unijunction transistor, in its first crude form, was patented in 1948 in France by Heinrich Welker, and the first commercial practical devices were produced in 1953 by the General Electric Company of the United States. Names varied: the company at first termed them *double base diodes*. Today they are called unijunction transistors, unijunctions or simply *UJTs*, which will be the term used in this book.

A UJT consists of a piece of semiconductor material, usually N-type, to which two connections are made. To this piece of N-type material is fused a pellet of P-type material, such that a PN junction is formed closer to one end than the other. A connection is then made to the P-type pellet. The N-type piece (the base) is only lightly doped, while the P-type pellet (the emitter) is heavily doped. This, together with the symbol for a UJT, is shown in Figure 15.1. The base end connection closer to the P-type pellet is termed 'base 2' and the other end 'base 1'.

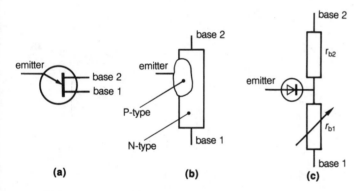

Fig. 15.1 *(a) Symbol, (b) construction and (c) equivalent circuit for a UJT*

Although basically a PN diode, the UJT operates in an entirely different manner. If the emitter and each base end are considered separately, the device would act as a diode—but it is not in fact at all employed in this fashion.

Consider the equivalent circuit for the UJT in Figure 15.1(c). This is shown as two resistors and a diode. The lower resistor, r_{b1}, is shown as a variable resistor and the upper resistor, r_{b2}, is fixed. The diode represents the PN junction formed by the P-type emitter and N-type base. The resistance between the two base connections, B_2 and B_1, is known as the *interbase resistance*, r_{bb}, and in the equivalent circuit could be written as,

$$r_{bb} = r_{b2} + r_{b1}$$

Now if the equivalent circuit is connected as shown in Figure 15.2, and the resistance of r_{b1} is made a maximum, very little current will flow, as the interbase resistance of a UJT can be from about 5 kilohms to 10 kilohms. Assume a positive potential (less than V_{BB}) is applied to the emitter terminal but, because of the voltage drop across r_{b1}, the diode is reverse biased and does not conduct. Now, as the resistance of r_{b1} is gradually reduced, a point will be reached when the voltage across the diode will be zero. Further reduction in the resistance of r_{b1} will reach a point when the voltage at the emitter terminal is more positive than 0.6 volt (the inbuilt PN potential barrier), and the diode will then conduct.

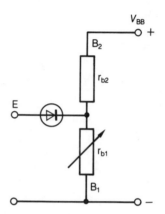

Fig. 15.2 *Action of the equivalent circuit for a UJT*

The operation of a UJT is similar to the above, but the change in resistance in the lower part of the base is brought about by the intrinsic behaviour of the UJT—not by external adjustment. If the UJT is connected in a circuit, as in Figure 15.3, with switch S_1 open, only a very small current will flow because (as has been stated) the interbase resistance is quite high since the N-type base is only lightly doped. In fact it behaves exactly like

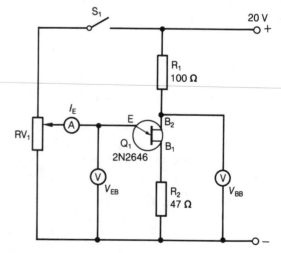

Fig. 15.3 *Test circuit for a UJT*

a simple resistor. This small current produces a voltage drop across the base such that the emitter is effectively reverse biased.

Now if switch S_1 is closed and the emitter voltage, V_{EB}, increased from zero, no significant current flows until a point is reached when the voltage at the emitter becomes greater than the internal reverse bias (provided by the base current) *plus* the inbuilt potential barrier. At this point the emitter conducts, reducing the resistance of the base. The emitter-base 1 voltage required to produce this conduction is termed the *peak point voltage* (usually given the symbol V_P).

What happens at the peak point voltage is that the effective emitter-base diode becomes forward biased. This allows holes to cross over from the P-type emitter zone into the N-type base. These holes in the region of the emitter and base 1 greatly increase the conductivity of the area. Greater conductivity means lower resistance and, because this lower resistance occurs with increasing current, it can be said that the UJT exhibits a *negative resistance* under these conditions. (There is no such actual thing as a negative resistance but this is the *effect* exhibited.)

The effect of the emitter-base 1 conduction can be seen in the curve in Figure 15.4. As the emitter voltage is increased there is a reverse emitter current flowing (due to minority current carriers) which changes to a positive current at point A on the curve. It is at this point that the emitter voltage is exactly equal to the internal reverse bias provided by the base current. With a further increase in the emitter voltage (about 0.6 volt) the peak point voltage is reached and the forward biasing of the emitter-base 1 junction reduces the resistance of the region so the voltage drops as the current increases (the negative resistance region on the curve). This position then stabilises and the emitter base resistance remains constant as any increase in voltage increases the current, as in a normal diode.

The peak point voltage of the UJT varies in proportion to the interbase voltage, that is the voltage between base 2 and base 1, V_{BB}, and may be expressed as:

$$V_P = \eta V_{BB} + V_D \qquad (15.1)$$

where V_P is peak point voltage,
$\quad$ η is the intrinsic standoff ratio,
$\quad$ V_{BB} is the interbase voltage, and
$\quad$ V_D is the emitter-base diode voltage drop.

The quantity η (the Greek letter *eta*) is termed the *intrinsic standoff ratio*. The value of η could be from 0.5 to 0.83 for particular UJTs. The value of η is the ratio of the lower resistor, r_{b1}, to the base-to-base resistance, r_{bb}, in the equivalent circuit of Figure 15.2, and could be written as:

$$\eta = \frac{r_{b1}}{r_{bb}}$$

However these values are not able to be determined, so the value of η may be obtained from the test circuit in Figure 15.3 using the approximate equation:

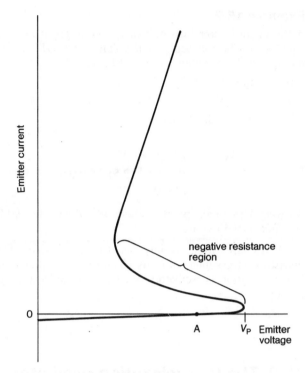

Fig. 15.4 *Characteristic curve of a UJT*

$$\eta \simeq \frac{V_{EB}}{V_{BB}}$$

where η is the intrinsic standoff ratio,
$\quad$ V_{BB} is the voltage between base 1 and base 2, and
$\quad$ V_{EB} is the voltage between emitter and base 1 at the moment of conduction, peak point voltage.

The value of η is given by the manufacturers of UJTs but quite often they will indicate a range into which the η of individual UJTs may fall. (For example a manufacturer may say η is between 0.65 and 0.75—a tolerance range.)

Example 15.1

Determine the approximate value of η for a UJT when, with a base-to-base voltage of 15 volts applied, the peak point voltage (when the emitter conducts) is 10.8 volts.

V_{BB} = 15 V
V_{EB} = 10.8 volts
η = ?

$$\eta = \frac{V_{EB}}{V_{BB}}$$
$$= \frac{10.8}{15}$$
$$= 0.72$$

The value of η for the UJT is 0.72.

Example 15.2

If the intrinsic standoff ratio of a UJT is given as 0.7, what would be the peak point voltage of a UJT when the base-to-base voltage is set at 12.5 volts?

η = 0.7

V_{BB} = 12.5 V

V_D = 0.6 (understood)

V_P = ?

$$V_P = \eta \, V_{BB} + V_D$$
$$= (0.7 \times 12.5) + 0.6$$
$$= 9.35 \text{ V}$$

Answer: The peak point voltage at which the UJT conducts is 9.35 volts.

UJTs are used mainly as trigger devices for thyristors—the subject of Units 18 and 19—but they also find application as oscillators (Section 15.2) and as timing circuits.

15.2 The UJT relaxation oscillator

If an RC circuit is combined with a UJT, a very interesting circuit evolves. With a given base-to-base voltage across the UJT, the peak point voltage will have a definite value which is set by the intrinsic standoff ratio of the particular UJT.

In Figure 15.5, the emitter of the UJT is connected across the capacitor of the RC combination. The emitter-base 1 voltage is therefore the capacitor voltage. When this voltage reaches the peak point voltage of the UJT, the emitter conducts and the current flows from the capacitor—through the emitter, base 1 and resistor R_3—to discharge the capacitor.

The RC combination on discharge (capacitor and combined resistance of emitter-base 1 and R_3) has a very small time constant. The resistance of R_3 is low and the forward resistance of the emitter-base 1 junction is also very low—so the time taken to discharge the capacitor is very short.

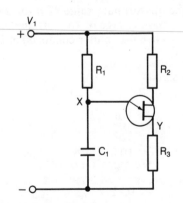

Fig. 15.5 *Unijunction relaxation oscillator*

As soon as the capacitor is discharged, conduction through the UJT ceases and the voltage begins to build up again, till peak point voltage is reached and the emitter conducts once more. This process is continuous and the complete circuit is called a *UJT relaxation oscillator.*

On Figure 15.5 two points in the circuit are noted. Point X, at the junction of R_1 and C_1, and point Y at base 1 of the UJT. At these points the waveforms of the UJT relaxation oscillator can be observed. Refer to Figure 15.6 to see these waveforms displayed. Both are called *sawtooth waveforms,* for obvious reasons. The waveform at point X (the input waveform or curve) is the charging of the capacitor through resistor R_1. As the resistance of R_1 could be quite high, say 20 kilohms, the rise in voltage across the capacitor is relatively slow. The waveform at point Y (the output waveform or curve) is of very short duration since the resistance of the emitter-base 1 and R_3 combination is small (as has been already stated), so the capacitor discharge time is very short. The waveform at point Y is more of a pulse than a sawtooth.

When the current through the emitter drops to a low value the resistance of the base rapidly rises and current ceases. This does not allow the capacitor to fully discharge, as can be seen in Figure 15.6.

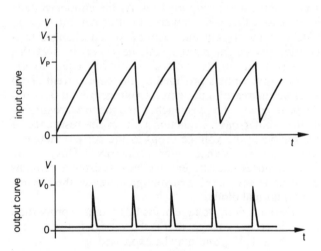

Fig. 15.6 *Input and output waveforms of a UJT relaxation oscillator*

The upper curve in Figure 15.6 is at the point X in Figure 15.5, and the lower curve at point Y. The upper curve is the capacitor charge and discharge curve. The capacitor charges up till the voltage across it reaches the peak point value. At this point the UJT emitter suddenly conducts and the capacitor is rapidly discharged. The lower curve shows that there is a very small residual voltage at point Y. This is caused by the small base-to-base current. At the instant the emitter of the UJT conducts there is a sudden rise in voltage drop across the base 1 resistor (R_3 in Figure 15.5), which rapidly falls to the original low value as the capacitor discharges. It is this pulse at point Y that is used to control the conduction in thyristors (discussed in Units 18 and 19). In thyristor trigger circuits the resistor, R_3, is usually replaced with a *pulse transformer.*

The period, or frequency, of the pulses delivered by the UJT relaxation oscillator can be varied by the value of the RC timing circuit, resistor R_1 and capacitor C_1. If the resistor is made variable, then the period of oscillation can be varied. An increase of resistance will increase the periodic time.

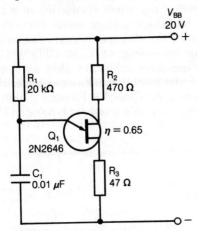

Fig. 15.7 *A practical UJT relaxation oscillator circuit*

The period of oscillation and the frequency can be determined from the approximate equation:

$$T \simeq 2.3\, RC \log \frac{1}{1 - \eta} \qquad \textbf{(15.2)}$$

where T is periodic time in seconds,
R is the timing resistor in ohms,
C is the timing capacitor in farads,
η is the intrinsic standoff ratio.

The frequency of operation is determined by taking the reciprocal of periodic time:

$$f = \frac{1}{T}$$

Consider the circuit of the unijunction relaxation oscillator in Figure 15.7. First let us find the voltage across the capacitor just before conduction (the peak point voltage). Using Equation 15.1:

$$\begin{aligned}
V_P &= \eta\, V_{BB} + V_D \\
&= (0.65 \times 20) + 0.6 \\
&= 13.6\ \text{V}
\end{aligned}$$

So the capacitor voltage will rise to 13.6 volts before emitter conduction takes place.

Now we will determine the approximate frequency of oscillation. Using Equation 15.2:

$$T \simeq 2.3\, RC \log \frac{1}{1 - \eta}$$

$$\simeq 2.3 \times 20 \times 10^3 \times 0.01 \times 10^{-6} \times \log \frac{1}{1 - \eta}$$

$$\simeq 0.00021\ \text{s}$$

So the approximate periodic time is 0.00021 second. Now, to find the approximate frequency of oscillation:

$$\begin{aligned}
f &= \frac{1}{T} \\
&= \frac{1}{0.00021} \\
&= 4762\ \text{Hz}
\end{aligned}$$

As the equation for periodic time is approximate (and remembering that the value of the resistor has a possible tolerance of 5 per cent and the capacitor probably 10 per cent) we can say the frequency is approximately 5 kilohertz. In practice it is usual to make R_1 a variable resistor and adjust the output to a desired frequency by using a CRO or frequency counter.

15.3 The PUT (programmable unijunction transistor)

The PUT is a four layer PNPN structure which is really a thyristor. Thyristors are covered in Unit 17, but the PUT will be dealt with in this unit as it is used as a thyristor trigger.

As the name (programmable unijunction transistor) suggests, the device acts as a UJT but with the added advantage that the standoff ratio (η) is not intrinsic, and is able to be programmed (or adjusted at will).

The symbol for a PUT, together with the four layer construction, is shown in Figure 15.8. The PUT is connected with two resistors, as in the circuit in Figure 15.9, and it is the ratio of the two resistors (R_1 and R_2) which determines the standoff ratio. The equation for determining the standoff ratio is:

$$\eta = \frac{R_1}{R_1 + R_2} \qquad \textbf{(15.3)}$$

where R_1 and R_2 are the voltage divider resistors,
η is the programmable standoff ratio.

The circuit in Figure 15.9 is for a PUT relaxation oscillator, which operates in a similar fashion to the circuit of Figure 15.7. For the circuit of Figure 15.9 let us determine the standoff ratio using equation 15.3.

$$\begin{aligned}
\eta &= \frac{R_1}{R_1 + R_2} \\
&= \frac{8.6 \times 10^3}{8.6 \times 10^3 + 10 \times 10^3} \\
&= 0.46
\end{aligned}$$

The standoff ratio of the PUT in the circuit is 0.46.

The PUT will conduct when the anode voltage is 0.6 volts above the anode gate voltage, i.e. when the anode-anode gate junction is forward biased. This is the peak point voltage for the circuit, and is derived from Equation 15.1, with the exception that V_1 replaces V_{BB}.

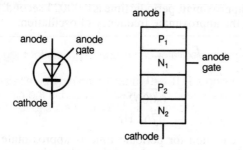

Fig. 15.8 *Symbol and layer construction of a PUT*

For the circuit in Figure 15.9 the peak point voltage would be:

$$V_P = \eta V_1 + V_D$$
$$= (0.46 \times 20) + 0.6$$
$$= 9.8 \text{ V}$$

The peak point voltage is 9.8 volts.

The frequency of operation for the circuit in Figure 15.9 may be determined from Equation 15.2 and the periodic time/frequency inversion. For this circuit it is approximately 1.7 kilohertz. Students may confirm this figure if they wish.

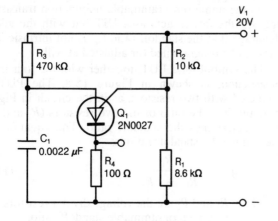

Fig. 15.9 *A PUT relaxation oscillator circuit*

15.4 The diac

The name *diac* is given to a two terminal device which will conduct in either direction when a certain *breakover voltage* (V_{BR}) is reached. The name 'diac' comes from *di*ode and *ac*. A diac may be of transistor (three layer) or thyristor (four layer) construction. In addition some diacs are actually ICs. No matter what type of

construction is employed, however, their operation is substantially the same.

As can be seen in the curves in Figure 15.10 the diac will operate in both directions. Whenever the voltage across the two terminals (simply referred to as terminal 1 and terminal 2) reaches a certain level the diac will change rapidly from a non-conducting to a conducting state. This breakover voltage varies from about 28 to about 36 volts. For this reason diacs cannot operate in circuits with low voltages (as can UJTs and PUTs). In addition, some diacs are deliberately assymetrical, i.e. the forward and reverse breakover voltages differ. It is usual to employ the symbol in Figure 15.10 for a diac—even though it may be a transistor type or an IC.

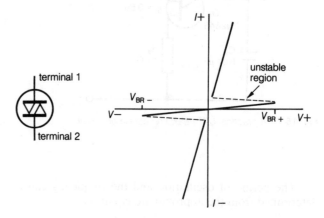

Fig. 15.10 *A diac symbol and operating characteristics*

Diacs will operate as relaxation oscillators similar to UJTs and PUTs, except that they operate at higher voltages. Because of their higher breakover voltage they are often employed directly to trigger SCRs and triacs, and this will be covered in later units. Figure 15.11 shows the circuit of a diac operating as a relaxation oscillator.

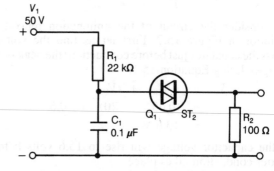

Fig. 15.11 *A diac relaxation oscillator circuit*

Unit 15 **SUMMARY**

- A UJT has two base connections, B_1 and B_2, and an emitter.
- A UJT has a high interbase resistance but when the voltage at the emitter rises to the peak point value the emitter suddenly conducts and the emitter-base resistance becomes very small.
- The peak point voltage is determined by the intrinsic standoff ratio of the UJT.
- A UJT can operate as a relaxation oscillator. The period of oscillation is determined by the value of an RC combination.

- A relaxation oscillator can produce pulses for triggering SCRs.
- When using a PUT the standoff ratio may be altered by choosing the ratio of two resistors used in the circuit.
- A diac is a two terminal device which is used to trigger SCRs and triacs.

Unit 16

Thyristors 1—the silicon controlled rectifier (SCR)

16.1 The thyristor family, and history of development

The name *silicon controlled rectifier*, usually shortened to SCR, was given to the device first developed in the laboratories of the General Electric Company in 1957 by Gordon Hall. Since the development of the first SCRs, other similar devices performing slightly different operations have been perfected. The name given to this class of device is *thyristor*, from *thyra*, Greek for 'gate', and *transistor*. Although the name may imply a three-terminal device, it also covers two- and four-terminal devices which may have unidirectional (i.e. the rectifier) or bidirectional conducting abilities.

The International Electrotechnological Committee (IEC) definition of a thyristor is: *a bistable semiconductor device, comprising three or more junctions, which can be switched from the off-state to the on-state or vice versa.*

Thyristors have two main functions: to act as controlling devices, and to act as *triggers* for controlling devices. Figure 16.1 shows the name, construction, conducting direction, and voltage and current ratings of thyristors used as controlling and trigger devices.

16.2 The SCR

The PN diode is a two-layer device, the BJT a three-layer device (either PNP or NPN), and the SCR is a four-layer device, usually referred to as PNPN. There are various methods of manufacture of SCRs but in essence the end construction is similar.

The four layers of the SCR, together with its symbol, are shown in Figure 16.2. Note that the third electrode is called the *gate*, while the other two, the anode and cathode, are similar in name and function to the PN diode.

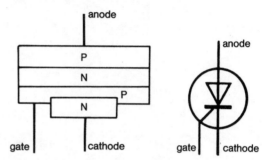

Fig. 16.2 *Section and symbol of a silicon controlled rectifier*

In the larger SCRS, all four layers are often sandwiched between tungsten backing plates which are in turn hard-soldered to a copper stud (in the case of the anode) and to a copper braid (in the case of the cathode). The gate lead must be just large enough to

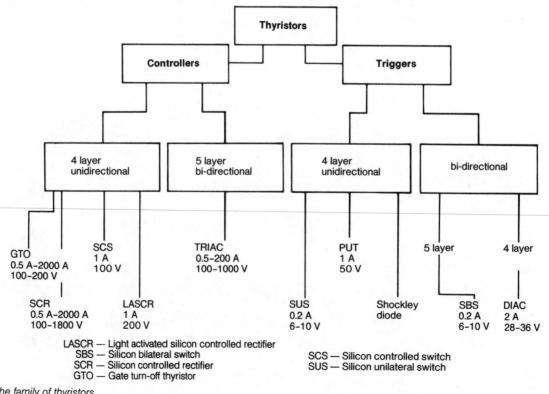

Fig. 16.1 *The family of thyristors*

be mechanically robust, as it only carries very small magnitudes of current.

The mounting stud is usually attached to a heatsink, so the heatsink is also the anode terminal. This is the reverse of most PN diodes where, in the larger sizes, the stud is the cathode. However, just as reverse diodes are made, so too are forward SCRs. The term *forward* and *reverse* simply mean the connections to the devices are reversed from the normal mounting arrangements.

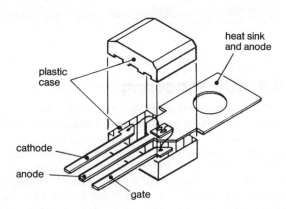

Fig. 16.3 *Flag heatsink SCR*

Smaller SCRs, say with a rating of 0.5 amperes at 200 volts, are packaged in the same way as small BJTs. Another common form for small SCRs (say 2 to 5 amperes up to 400 volts) is as a small rectangular plastic encapsulation, with a flag terminal for the anode which can act as a heatsink (Fig. 16.3). This is usually referred to as the TO-220 package. These, together with stud mounting and screw mounting types of SCRs, may be seen in Figure 16.4.

16.3 Principle of operation

Although detailed exploration of the characteristics of SCR operation will not be discussed in this book, a brief explanation will assist students and readers.

An SCR is made up with two PN junctions, from the PNPN construction (Fig. 16.5). The gate and cathode lead can be considered to be between the lower PN pair. Now if the gate has a positive potential, with respect to cathode, the junction of the lower PN pair is forward biased and acts like a closed switch. If, at the same time, a positive potential is applied to the anode, the junction of the top PN pair is forward biased and will also conduct. Now since both PN pairs are forward biased there will be current from the anode to cathode. The current will provide the correct voltage drops to keep the two PN pairs forward biased. This will continue even if the gate current is removed. Under these conditions the SCR will conduct as long as there is sufficient current through the device.

If the positive potential is removed from the gate, it has no effect on the conducting ability of the SCR, since the positive anode potential is carried through the device. When the positive potential is removed from the anode, conduction ceases and the SCR reverts to its non-conducting state—even after the positive potential is reapplied to the anode. If a positive potential is applied to the anode, and negative to the cathode (before any gate current exists), no current will flow past the top PN pair since, without any gate current, the lower PN pair is not forward biased.

Conduction may be restarted either by applying a positive current pulse to the gate or increasing the anode potential to the *breakover point*. When this breakover occurs, the positive potential is carried through to the lower PN pair. This again forward biases the lower PN pair and the SCR goes into a conducting state. It is not

Fig. 16.4 *A range of SCRs from 5 amperes to 140 amperes. Voltages range from 500 V to 1600 V PIV*   PHILIPS ELECTRICAL COMPONENTS AND MATERIALS

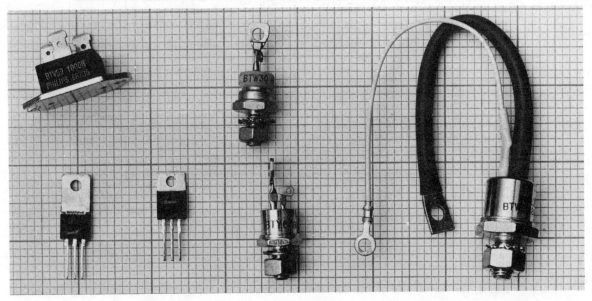

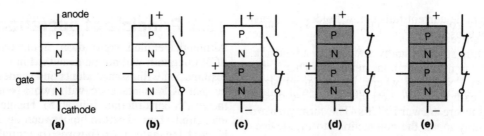

Fig. 16.5 *Operation of an SCR: (a) An SCR consists of two PN junction pairs (b) Positive at anode, negative at cathode — no conduction. The two PN pairs are like open switches (c) Gate positive, cathode negative — lower PN pair is forward biased and acts like a closed switch (d) Anode and gate both positive, cathode negative — both PN pairs are forward biased and act like closed switches (e) After conduction is initiated the current from anode to cathode keeps the two PN pairs forward biased and the positive potential can be removed from the gate. Both PN pairs act like closed switches*

usual to operate an SCR in the breakover mode—it is usually brought to a conducting state, or *fired*, by gate current. The characteristic curves of an SCR in Figure 16.6 illustrate these two firing conditions.

Note that there are two loci (paths) to the forward curve. One, (*a*), in which no gate current is flowing, and the other, (*b*), in which gate current allows a forward characteristic similar to a PN diode. (Lower values of gate current will produce curves intermediate to these, but are omitted for reasons of clarity.) Note that if the forward voltage is kept below the breakover point no real conduction can take place until gate current flows. When this happens anode current rises to the normal conduction region. A further point on the curve is marked *holding current*: this is the minimum current that will allow the SCR to remain in the conducting condition; if current falls below this value the SCR will revert to the non-conducting condition.

16.4 SCR triggering

An SCR must be triggered into conduction by applying a small current to the gate terminal. The SCR is a current-operated device and requires this current to enable triggering to take place. Both the magnitude and duration of the current are important in obtaining correct triggering. The maximum gate-to-cathode voltage is usually quite small—a typical value would be 5 volts, and a typical operating voltage 2 volts. The gate current required for conduction would be about 35 milliamperes for a period of from 10 to 50 microseconds.

When current flows in the gate of the SCR to enable triggering to take place, it is not necessary for it to remain once the SCR *latches*. The circuit in Figure 16.7 illustrates the procedure. When push button PB1 is depressed the SCR will commence to conduct. When it is released conduction will still take place. Now, it has been said

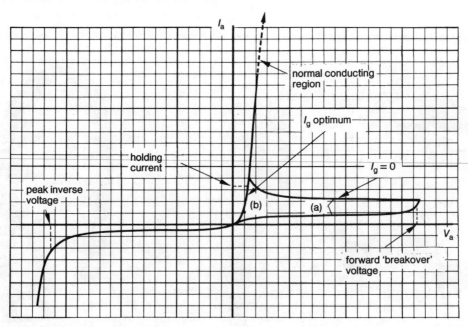

Fig. 16.6 *Characteristics of a silicon controlled rectifier*

that when the anode current falls below holding current, conduction will cease: in Figure 16.7 this would be done by pressing push button PB2. This will open the anode circuit, current will fall to zero (certainly below holding current) and the SCR thus reverts to a non-conducting state. Alternate operation of PB1 and PB2 will cause the SCR, and lamp L, to be alternately on and off.

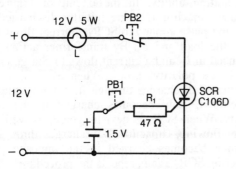

Fig. 16.7   *Simple commutation by current interruption*

The switching off of an SCR is referred to as *commutation*—from the Latin, and simply meaning a complete, or thorough, change, i.e. from a conducting to a non-conducting state. The most obvious method of commutation is that shown in Figure 16.7: opening the anode circuit. However two other simple methods are also sometimes used. These are illustrated in Figure 16.8, and consist of shorting out the SCR, or placing a capacitor in parallel with it. The short circuiting switch simply bypasses the SCR, reducing its current to below holding value; and the capacitor, on charging, also momentarily reduces the anode current to below holding current value. Both methods will commutate the SCR provided no gate current is flowing. All three methods are referred to as *natural commutation*.

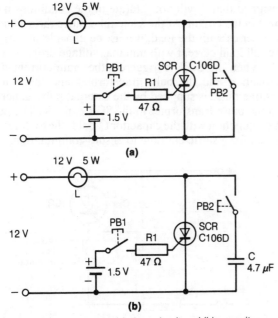

Fig. 16.8   *Commutation by (a) short circuit and (b) capacitor*

For an SCR to go into conduction, the forward current must be a certain minimum value while triggering is taking place. This value could be about 300 per cent of holding current and is termed *latching current*. In this respect the SCR is similar to a series-shunt relay. A certain current is required to bring it into operation, and a minimum current must flow to enable it to remain energised. This is shown in Figure 16.9 and is often referred to as the *SCR series-shunt relay analogy*. A momentary positive potential applied to the 'gate' terminal will cause the shunt coil to energise the contact, and current will flow from the 'anode' to the 'cathode'. This load current, if sufficient, will allow the series coil to hold in the contact and the potential can be removed from the 'gate'. If the current falls to too low a value the series coil can no longer hold in the contact and it will drop out to open the circuit.

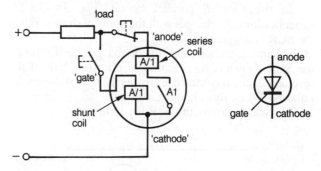

Fig. 16.9   *The series-shunt analogy of an SCR, with equivalent circuit on left and SCR symbol on right*

16.5 Forced commutation

If natural commutation cannot be used—and this is usually the case—then an SCR must be commutated by *forced commutation*. This means attempting to reverse the current through the SCR by some external means. There are six such classifications:

Class A self-commutated by resonating the load;
Class B self-commutated by parallel LC circuit;
Class C capacitor or LC switching by another load-carrying SCR;
Class D capacitor or LC switching by an auxiliary SCR;
Class E commutation by an external pulse, and
Class F ac line commutation.

We will now examine each of these in turn. (In Figures 16.10 to 16.15 the triggering circuits have been omitted for clarity, since only the on-to-off state commutation is of interest.)

Class A commutation is obtained by placing an inductor and capacitor, with a damping resistor in parallel, in series with the load and the SCR. When the SCR is triggered the resulting load current excites the LC circuit, which commences to resonate. After half a resonant cycle the LC circuit starts to reverse the current

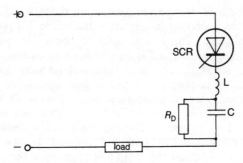

Fig. 16.10 *Class A self commutation for SCRs*

and so commutates the SCR. It is evident that for a continuous average forward current to flow, the SCR would have to be continually triggered. The circuit for this class of commutation is shown in Figure 16.10.

In the *Class B* parallel circuit (Fig. 16.11) the capacitor initially charges with the polarity shown. When the SCR is triggered, current flows through the load and at the same time the capacitor discharges itself through the SCR. When it has discharged after one half of the resonant cycle, it commences to charge with the opposite polarity. When this charging current is greater than the SCR forward current, the SCR is commutated.

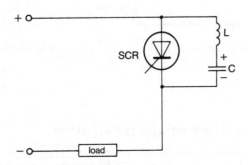

Fig. 16.11 *Class B self commutation for SCRs*

Class C commutation utilises two load-carrying SCRs. In Figure 16.12, if SCR$_2$ is conducting, capacitor C will be charged with the polarity shown. When SCR$_1$ is triggered, C is switched across SCR$_2$ via SCR$_1$. The discharge current of capacitor C opposes the main current in SCR$_2$ and commutates it. If SCR$_2$ is triggered, the process may be repeated.

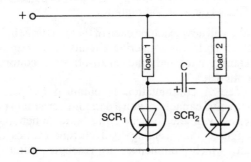

Fig. 16.12 *Class C commutation for SCRs*

A *Class D* circuit could be made from the circuit in Figure 16.12 by making one SCR a low power auxiliary unit with a value of load resistance such that only a little over holding current is maintained in the auxiliary SCR. However, the best example of the Class D circuit is the Jones Chopper. This circuit is given in Figure 16.13.

The Jones Chopper circuit has an extremely reliable commutating ability. In the circuit of Figure 16.13 consider capacitor C to be initially discharged and a triggering pulse applied to SCR$_1$. Current flows through SCR$_1$, the load and L_1. By transformer action an emf is induced in L_2 and a current flows to charge capacitor C with the polarity shown. When SCR$_2$ is triggered, capacitor C discharges through it, commutating SCR$_1$. Capacitor C now becomes charged in the opposite polarity. When SCR$_1$ is next triggered, as well as load current flowing, capacitor C discharges through SCR$_1$ and then becomes charged in the opposite polarity. Triggering SCR$_2$ again, repeats the procedure.

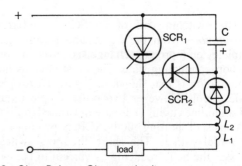

Fig. 16.13 *Class D Jones Chopper circuit*

Class E commutation utilises an external source to pulse the load carrying SCR into commutation. Figure 16.14 shows a suitable circuit for this type of commutation. It is important that the transformer is so designed that it will not saturate in the conditions it will meet in the circuit. At the same time, since the secondary is in series with the load, it must be capable of carrying the full load current with minimal voltage drop.

When the SCR is triggered, the main current flows through it and the pulse transformer and the load. A positive pulse is supplied by the external pulse generator via the pulse transformer to the SCR cathode. The pulse, in conjunction with the capacitor C, holds the SCR reverse biased for a sufficient time to ensure commutation.

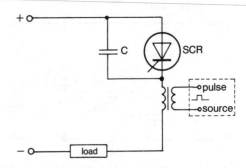

Fig. 16.14 *Class E commutation of an SCR*

Class F commutation only applies when the SCR is connected to an alternating supply. In Figure 16.15 when the anode is positive, during a positive half-cycle, the SCR is triggered. Conduction takes place until the end of the positive half-cycle, when the SCR is commutated as the negative half-cycle commences.

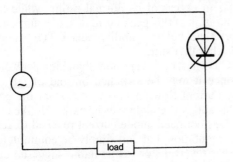

Fig. 16.15   *Class F commutation of an SCR by ac line*

16.6 SCR applications

SCRs are ideal for *static switching*, i.e. the switching of a power circuit having no moving parts. This may be for silent operation, greater ease of control, or operation where intrinsic safe circuits are required, e.g. in explosive atmospheres. A simple switching circuit is shown in Figure 16.16. When switch S_1 is closed, the SCR will be triggered at about the start of the positive half-cycle. When the SCR conducts, almost the full voltage of the supply will appear across the load, and so gate current will cease.

When the ac wave reaches zero (at the end of the positive half-cycle), the SCR is commutated (class F) and conduction ceases. On the negative half-cycle no current flows—so this circuit is a half-wave controlled circuit only.

Diode D_1 prevents reverse current flowing in the gate during the negative half-cycle, and resistor R_1 limits the value of gate current before conduction takes place.

As long as switch S_1 is closed, the SCR will be triggered at the start of each positive half-cycle and will switch off (become commutated) at the end of each positive half-cycle.

The half-wave limitations of the circuit in Figure 16.16 may be overcome with the full-wave circuit shown

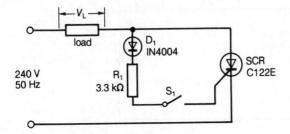

Fig. 16.16   *Half-wave SCR static switch*

in Figure 16.17. In this circuit SCR_2 is triggered at the start of each positive half cycle, and SCR_1 at the start of each negative half-cycle, as long as switch S_1 is closed. Thus each SCR conducts for 180^0 of the complete cycle and each is commutated at the end of its conducting half-cycle. In Unit 17 it will be seen that the two SCRs may be replaced with a single triac.

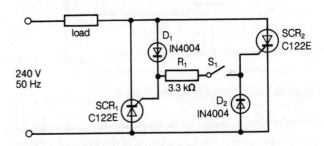

Fig. 16.17   *An ac static switch*

Static switching may also be used on direct current circuits where the SCR may act as a latching contactor. This operation is shown in Figure 16.18, where Class D commutation is used. When push button PB_1 (the 'on' button) is pressed, SCR_1 is triggered and conducts. Current flows through the load and SCR_1 and charges the capacitor C with the polarity shown. Now when PB_2 (the 'off' button) is pressed, SCR_2 is triggered and commutates SCR_1 by 'dumping' the charged capacitor across it. SCR_2 will remain conducting with current only just above holding current flowing, limited by resistor R_1. Capacitor C will now be charged in the opposite polarity and will commutate SCR_2 when SCR_1 is next triggered by pressing PB_1.

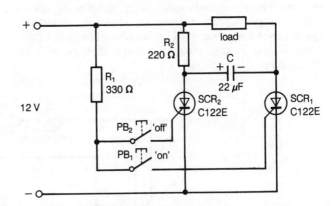

Fig. 16.18   *A push-button SCR 'contactor'. Forced commutation by an auxiliary SCR*

The circuit of Figure 16.19 employs an SCR as a latching relay. In this circuit if any of the contacts are momentarily closed, current flows into the gate of the SCR and it is triggered into conduction. The SCR will remain conducting, even if all the contacts re-open, and will indicate an alarm condition until the reset button is pressed and the SCR is commutated by natural commutation.

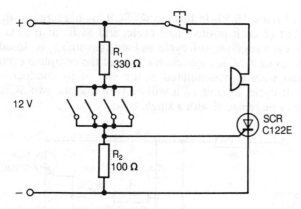

Fig. 16.19 *An SCR latching-signal circuit*

The circuit of Figure 16.20 represents an emergency lighting system which will activate in the event of a mains failure. This sort of system is required in theatres and public halls and is often referred to as *panic lighting*. The 12 volt battery is trickle-charged through diode D_1 and resistor R_3 as a half-wave rectifier. The capacitor C_1 is charged through diode D_2 with the polarity shown. This places a negative potential at the gate of the SCR so it is not triggered.

In the event of a mains failure, capacitor C_1 is reverse charged (by the battery) through R_2 and the transformer secondary winding. When the gate is positive enough the SCR is triggered and current flows from the battery, through the lamps, the SCR and the transformer secondary-winding back to the battery. When the mains are re-energised the SCR is commutated by reverse polarity at the cathode and the lights will go out.

The above applications of SCRs as static switches are only a very small sample of the very wide range of uses. Others include time delay circuits, regulated power supplies, motor controls, choppers, inverters, cycloconverters, protective circuits, phase controls, heater controls and battery chargers. (Some of these will be discussed in Units 18 and 19.)

16.7 Gate turn off SCRs (GTOs)

The *gate turn off SCR*, also known as a gate turn off switch, is more commonly referred to simply as a *GTO*. The GTO is a four-layer thyristor, somewhat similar to an SCR, which can block or pass a current in the forward direction. It also has the ability to pass a forward peak current far in excess of its normal rating, while turned on. The duration of this peak current is, of course, limited to a short time but this ability puts GTOs somewhat ahead of other thyristors.

The GTO has, however, more than high peak current ability, since it may be switched on and off at a high speed, by current flowing into or out from the gate. It is from this ability that it derives its name. Unlike normal SCRs it does not need anode current reversal to achieve forced commutation, but can simply be commutated by gate current reversal—or (as its name suggests) can be 'turned off' by the gate.

In operation it behaves as both a BJT and SCR and has the most desirable features of both devices. The symbol for the GTO is shown in Figure 16.21. Note the two arrows at the gate-to-cathode connection, illustrating that current will flow in both directions.

The GTO is packaged similarly to SCRs and some other thyristors, and three styles are shown in Figure 16.22.

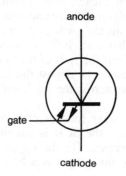

Fig. 16.21 *Symbol for a GTO*

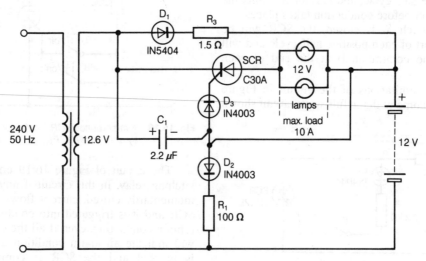

Fig. 16.20 *An SCR-activated emergency lighting circuit*

Fig. 16.22   *Three types of GTO*

16.8 GTO operation

Figure 16.23 shows the forward characteristics of a GTO. Note particularly that the curves are divided into two sections, one section being transistor operation and the other thyristor operation.

Until the gate current reaches a certain value (200 milliamperes in our case), the GTO behaves as a BJT, with a gate to anode current amplification factor which increases with anode current. When the gate current is less than required for triggering (in this case 200 milliamperes), only a low current (controlled by gate current) flows between anode and cathode. The anode-cathode voltage drop may be quite high (depending on the particular GTO characteristics) in this state. When the gate current reaches the triggering value, usually referred to as *gate trigger current*, I_{GT} (again about 200 milliamperes in the curves of Figure 16.23), the GTO is put into the 'on-state' with a small potential difference between anode and cathode.

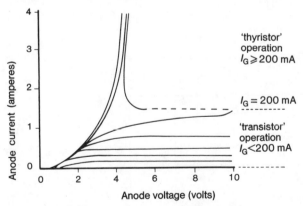

Fig. 16.23   *Anode characteristics of a GTO for various levels of gate current*

If the anode current is below latching current, the GTO may revert to the 'off-state' with removal of the gate current. However, if the anode current is above latching current, the gate current may be removed and anode conduction will continue.

If the gate current is reversed (i.e. the current is made to flow from the gate), unlike an SCR the GTO will be turned off.

The circuit in Figure 16.24 represents the gate control circuit for a GTO. This is known as *direct gate drive* as it is not isolated from the GTO. In other applications, especially when the GTO is controlling high voltages, an *isolated gate drive* circuit would be employed. From Figure 16.24 it can be seen that the gate of the GTO is turned on when the current is supplied through the conducting Q_1. Initially when Q_1 conducts, the charging current flows into C_1 and via L_1 to the gate. This pulse of current gives good switching characteristics. After C_1 is charged and current ceases to flow, gate current is maintained via R_2 to ensure the GTO remains in conduction.

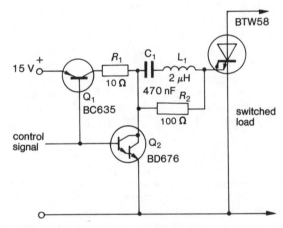

Fig. 16.24   *Simple gate drive circuit for GTO*

When the Darlington BJT, Q_2, is brought into conduction (with a control signal), the join between R_1 and C_1 goes negative and C_1 discharges and draws current from the gate of the GTO. This turns off the GTO.

The GTO is finding ever increasing applications, especially in dc motor speed control. This ranges from control of small motors in household appliances (such as washing machines) to stepless speed control on mainline electric trains. Other applications include dc to ac inversion—and units up to and over 30 kVa are in use. This application is used on dc supplied trains to provide ac for the train auxiliaries, such as air conditioning.

Unit 16 **SUMMARY**

- The SCR is one of a number of devices which are classified as thyristors.
- The SCR is a high power unidirectional controlled diode in which current flowing into the gate turns it on and reverse biasing turns it off.
- The process of applying a small pulse of current to the gate of an SCR and causing it to conduct, is called triggering.
- Opening the anode circuit or attempting to reverse bias the SCR and causing it to stop conducting, is called commutation.

- Reducing the anode current below holding current is called natural commutation.
- Attempting to reverse the current through an SCR is called forced commutation.
- SCRs are extensively used in industry—one application being their use as static switches.
- GTOs are thyristors which can be turned both on and off by gate current.

Thyristors 2—the triac

17.1 Triac characteristics

In Section 16.7 it was mentioned that the two SCRs in Figure 16.17 could be replaced with a *triac*. A triac acts like two SCRs connected in inverse parallel and so could be termed a *bi-directional* or *two-way current carrying* controlled device. The name 'triac' comes from *tri* meaning three (terminals) and *ac*, meaning it will operate on alternating supplies.

A triac is classed as a five-layer semiconductor device but reference to Figure 17.1 will show it has six separate sections—four of which are made from N-type material and two of P-type. Both the construction and theory of triac operation are rather complex and will not be fully discussed.

The triac symbol also appears in Figure 17.1 and it can be seen that it is similar to the diac symbol except that it has a 'gate' terminal. As it is bi-directional, the terms 'anode' and 'cathode' cannot be applied so the two power terminals are simply referred to as *terminal 1* (or T_1) and *terminal 2* (or T_2). The gate performs the same function as an SCR and can trigger the triac into the conducting mode in either direction.

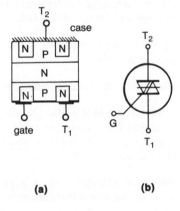

(a) **(b)**

Fig. 17.1 *(a) Section and (b) symbol of a triac*

The characteristic curves of the triac are shown in Figure 17.2 and it can be seen that in the positive direction (in the quadrant marked T_2+) the operation of the triac is identical to that of the SCR. However, as it is bi-directional, conduction will also take place in the reverse direction, and this conducting mode can be seen in the third quadrant (marked T_1+). In both these quadrants of the graph it can be seen that there is a *breakover voltage* on the x axis (marked V_{BR}), similar to the diac characteristics in Figure 15.10—but, like the SCR, the triac is seldom operated in this condition. As the triac will conduct in either direction there is no such term as peak inverse voltage applied to the triac. Triacs are

packaged similarly to SCRs and could be mistaken for SCRs (or vice-versa) unless the type number was recognised. Also, in some instances an SCR could be replaced by a triac—but a triac could not be replaced by a single SCR.

17.2 Triac operation

In normal operation the triac is operated well within the breakover voltage rating and is triggered by a pulse from the gate. If gate current flows at the beginning of each half-cycle, the triac will conduct through the dashed line in the curves in Figure 17.2, and current will flow in either direction with only a very small voltage drop and very small forward resistance.

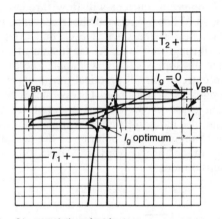

Fig. 17.2 *Characteristics of a triac*

Like the SCR the triac too has a minimum current to maintain conduction, again termed *holding current*, and another minimum current, *latching current* (somewhat greater than holding current) that must flow before gate current can be removed.

17.3 Operating modes

There are four operating modes for triacs. (By *operating mode* we mean the type of gate current required to initiate conduction in either direction.) In a triac, gate current may flow into or out from the gate. This means that either a positive *or* a negative potential applied to the gate will initiate conduction. As conduction takes place in the two quadrants of the graph in Figure 17.2 and as these are usually referred to as quadrant I and quadrant III of the voltage/current (*VI*) characteristics, we use the quadrant roman numeral (I or III) to indicate the two operating directions.

In quadrant I (when terminal T_2 is positive) if the gate is also positive we say the operating mode is I+. If T_2 is positive and the gate is negative we call this the I- operating mode. Now when T_1 is positive and the gate is positive the operating mode is III+. When T_1 is positive and the gate is negative this is called operating mode III−.

The preferred modes of operation are I+ and III−. The sensitivity to triggering is much reduced in mode I− and even more so in III+. This does not mean that the other modes are never used. In some circuits gate current may not be able to be changed, but if at all possible the I+ and III− modes are employed. It must also be stated that some triacs will not operate in mode III+.

17.4 Simple triac circuits

Like the SCR the triac can also be used as a static switch. Figure 17.3 shows the circuit of a simple static switch which will initiate conduction in either direction (or both halves of the ac cycle) when switch S_1 is closed. With switch S_1 open, no current will flow through the triac and the load. When S_1 is closed, the triac will be triggered into conduction at the start of each half-cycle, and ac will flow through the load. At the instant of triggering, almost the full supply voltage will appear across the load and gate current (for all practical purposes) will cease, even though S_1 was still closed. On opening S_1 conduction will again cease as the triac is commutated (class F) at the end of each half-cycle. It will also be noted that, with the changing of the polarity of the ac supply, the operating modes are I+ and III−. These, of course, are the preferred modes.

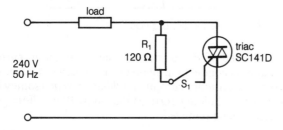

Fig. 17.3 *A simple triac static switch*

A further ac switching circuit is shown in Figure 17.4. Here the gate triggering is operated by the transformer, *T*. When the transformer secondary switch S_1 is open, the high impedance of the transformer primary winding drops almost all of the supply voltage across it, and triggering will not take place. When switch S_1 is closed the transformer winding impedance falls to a very low figure and triggering will take place at the beginning of each half-cycle, similar to the action in Figure 17.3. Because only extra low voltage appears at the terminals of switch S_1, this circuit would be useful where higher voltages could be risky, such as in a very damp or very conductive situation.

Triacs, as their name implies, are usually only used on ac circuits, both single-phase and three-phase. They have very many applications apart from use as static

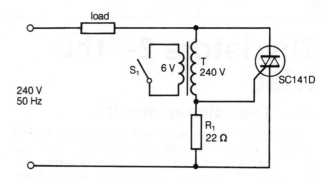

Fig. 17.4 *A simple triac static switch using low voltage control*

switches: light dimming, motor speed control, motor starting, furnace control and heating control, for instance. Some of these applications will be examined in Unit 18.

17.5 Thyristor terminology

The following is a glossary of some terms connected with the thyristors covered in Units 15, 16 and this unit.

Thyristor Collective name for four-layer (or more) devices which may be switched to on-state or off-state or vice versa.

SCR Silicon controlled rectifier—a rectifier with a control terminal, called a gate, which determines at what anode-to-cathode voltage the device will conduct.

Triac The equivalent of two inverse-parallel SCRs with a common gate terminal.

Diac Similar to a triac but without any gate terminal.

PIV Peak inverse voltage—also known as reverse breakdown voltage. The maximum reverse voltage that may be applied to an SCR before it breaks down (and is permanently ruined).

Forward breakover voltage The voltage at which a thyristor will switch on when the gate current is zero.

Triggering Initiating conduction in a thyristor by either breakover voltage or by gate current.

Average forward current The maximum permissible average current that can be safely carried by an SCR.

RMS current The maximum permissible rms current that may be safely carried by a triac.

Latching current The minimum current required for a thyristor to remain in conduction at the instant of triggering.

Holding current The minimum current for a thyristor to remain in the conducting state with no gate current flowing.

Unit 17 **SUMMARY**

- A triac is a controlled two-way current carrying device.
- The three terminals of a triac are called terminal 1, terminal 2 and gate.
- Triacs are similar to SCRs in appearance.

- The best operating modes for triacs are I+ and III−.
- Triacs can be used as static ac switches.
- Triacs are widely employed in the control of ac circuits.

Thyristor ac voltage control

18.1 Phase control

In Unit 15 triggering devices for thyristors (viz. UJTs and diacs) were discussed, with the emphasis on the fact that the rate at which each device could produce a pulse of triggering current could be varied. The UJT produces pulses in a single direction but the diac is able to produce pulses in either direction, as it is a bi-directional device.

In this unit we will examine the use of triggering devices in controlling the firing of thyristor circuits. In the circuits of Units 16 and 17 only basic triggering was discussed since the thyristor devices were simply used as static switches. While this use of thyristors has importance, by far the most useful application is the controlling of a supply to achieve voltage and current variation for a variety of situations. These could include motor speed, heating and lighting control.

Thyristors are usually controlled by *phase control*. Phase control of thyristors means allowing conduction for only part of an ac cycle. It naturally follows that for this type of control to be used the primary supply must be alternating. It is possible to trigger an SCR from any point from 0^0 to 90^0 in a half-cycle with simple resistor triggering, as shown in Figure 18.1. Here the gate current is set by varying resistor R. When the anode voltage reaches a certain level (depending on the magnitude of the gate current) conduction will start. It can be seen from the curves that by altering resistor R from zero to maximum resistance, conduction can be initiated from any point on the ac half-cycle, from the commencement to maximum value. The SCR is commutated at the cycle end. As the maximum value is as high as the anode voltage can get, any attempt to further reduce the current by increasing the resistance will result in no firing of the thyristor and no current flowing at all. The diode, D, prevents reverse gate current flowing on the negative half-cycle.

The simple resistive control as shown in Figure 18.1 is seldom very satisfactory for a number of reasons. One is that we can only control the output current up to 90^0 in the half-cycle and, in addition, the actual angle at which conduction commences is rather indeterminate and can vary greatly with slight changes in supply voltage. For this reason it is seldom used. However it is a form of phase control, as the *trigger angle* (the angle between zero and when triggering takes place), has a time relationship, or phase difference, with the input voltage wave.

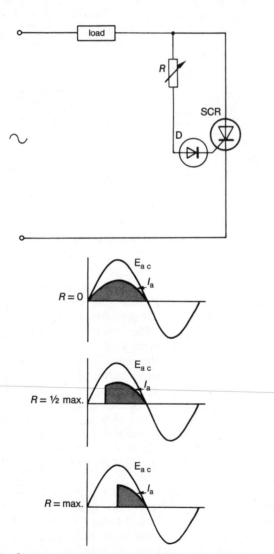

Fig. 18.1 *Simple resistor triggering of an SCR on alternating supply*

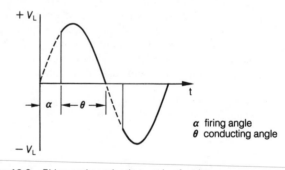

Fig. 18.2 *Firing and conducting angle of a triac*

Figure 18.2 illustrates the terms used in phase control. On the curve we can see two time intervals marked: α (the Greek lower case letter *alpha*), and θ (Greek lower case *theta*). These intervals may be described as:

Trigger angle, α — the point in the cycle when the trigger or gate pulse is applied, measured from the start of the cycle.

Conduction angle, θ — the period of the cycle for which the thyristor conducts.

18.2 RC phase control using a diac

In the circuit of Figure 18.3, if a current pulse can be applied to the gate of the thyristor (this time a triac) it could produce conduction at any point in both half-cycles of the ac wave. This would mean that the angle α could be anything from zero degrees to almost 180^0 and the conduction angle θ, anything from 180^0 to almost zero degrees. The effective load voltage, V_L, and therefore load current, I_L, in each half-cycle could be controlled from full value to almost zero. A complete circuit to achieve this effect can be seen in Figure 18.4.

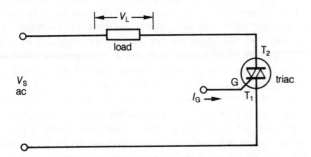

Fig. 18.3 *When a triac is triggered by gate current the full supply voltage appears across the load*

The circuit of Figure 18.4 is a simple triac light-dimming circuit. At the commencement of each half-cycle the capacitor C_1 charges through variable resistor R_1. When the voltage across the capacitor, V_C, reaches the breakover voltage of the diac (for the ST2 diac, about 30 volts) the triac triggers and conducts to the end of the half-cycle, when it is commutated Class F. This occurs on both the positive and negative half-cycle and, it can be seen, the triggering modes are I+ and III−. Now if the resistance of R_1 is increased so that it takes longer to charge capacitor C_1 it will take longer, in the half-cycle, for the diac breakover voltage to be reached. This means that with adjustment of R_1 the triac may be triggered at any angle, α, from just above zero degrees to almost 180^0.

This would mean that, in the circuit shown in Figure 18.4, triggering would not take place until α equals five

degrees. However, in a practical application, the eye would have great difficulty discerning the difference in lamp brightness that would result from triggering at zero degrees or five degrees. In addition, for lamp dimming, the conduction angle, θ, would not be much less than 90^0, as at this setting the lamp would only be giving about 25 per cent of its maximum brightness. Further reduction in brightness below 25 per cent can be obtained (till light output is zero), but this does not give any useful illuminance. The result is an excellent lamp-dimming circuit in which power loss is minimal.

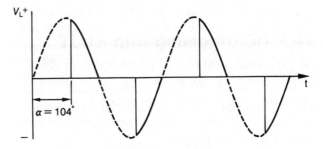

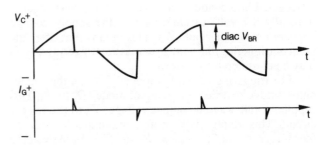

Fig. 18.5 *Waveforms for the circuit of Figure 18.4. At the top is load voltage; centre capacitor voltage; and bottom gate current*

Figure 18.5 shows the operating conditions at a particular setting of R_1 in Figure 18.4. In this case R_1 is set so that diac breakover voltage, V_{BR}, does not take place until α is 104^0. The upper curve shows the load voltage curve (load current would be in the same form)— the dashed portion of the curve being the original supply voltage wave. The middle curve represents the voltage across the capacitor, V_C. The capacitor charges up until the voltage across it reaches the diac breakover voltage, V_{BR}. At this point the diac conducts, quickly discharging the capacitor and producing the current pulses shown in the lower curve. It is these pulses which trigger the triac into conduction in both the positive and negative half-cycles.

Figure 18.6 shows a triac replaced by two SCRs. The triac, you will remember, behaves as two inverse parallel SCRs, and replaces them. However, in very high power-circuits we are limited by the triac current ratings (see Figure 16.1) and so pairs of high current rated SCRs are used.

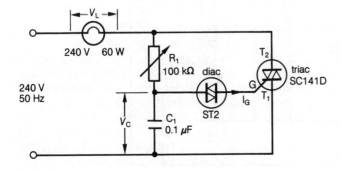

Fig. 18.4 *Simple phase control light dimming circuit*

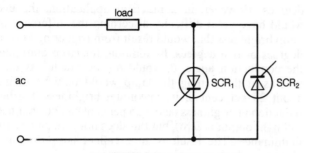

Fig. 18.6 *Inverse parallel SCRs for a larger current than a single triac could carry*

18.3 RC triggering with a UJT

As has been mentioned at the beginning of this Unit the UJT can also trigger thyristors. A control circuit using a triac which is triggered by a UJT is shown in Figure 18.7. This circuit contains many features previously discussed in this book and we will go through the circuit discussing each point.

The main part of the circuit is the power circuit which consists simply of the load and the triac. When the triac is triggered into conduction its voltage drop is very small (typically 0.7 volt) so almost all of the supply voltage appears across the load (V_L). This means that after the point of triggering, the voltage across the triggering circuit may be considered to be zero.

The triggering circuit is supplied through the single-phase bridge rectifier consisting of diodes D_1 to D_4. This supplies a full-wave dc output voltage, V_1, to the trigger circuit. This voltage is kept at a maximum voltage by the shunt regulator formed by R_2 and ZD. As the voltage increases, from the beginning of a half-cycle, it will rise

to a maximum value, V_Z, determined by the voltage of the zener diode ZD. The resistor R_3 and the capacitor C_1 form an RC timing circuit which—when combined with the UJT, resistor R_4 and the primary of the pulse transformer, T_1—form a UJT relaxation oscillator.

Figure 18.8 shows the waveforms in the triggering circuit of the triac control in Figure 18.7. At the commencement of a half-cycle, the rectified circuit voltage, V_1, rises to a point, V_Z, when the zener diode begins to conduct. Also, from the commencement of a half-cycle the capacitor begins to charge through R_2, and its voltage (V_C) rises until at the peak-point voltage (V_P) the UJT conducts and current flows through the primary of T_1. This is only a very short current pulse but it induces a voltage into the secondary winding of T_1 which is applied to the gate of the triac and triggers it into conduction. Note particularly that the secondary winding polarity mark (the dot) shows that the secondary connections have been reversed. The reason for this is that the current trigger pulses from the UJT can only be in one direction—so by providing negative pulses, the triac can be triggered in the I- and III- modes. If the pulses were positive, the triggering would be in I+ and III+ modes which are the least desirable (as some triacs will not operate in III+).

The use of a pulse transformer, in the circuit in Figure 18.7, is desirable for many reasons—two of which are that the control circuit can be electrically isolated from the triac gate, and the polarity of the output pulses can be altered (as is done here). Another reason is that, in the circuit of Figure 18.6, two secondaries could provide pulses of opposite polarity to the two SCR gates. A pulse transformer is quite a simple device and could consist of just a few turns of wire on a ferrite toroid (ring). The turns ratio of primary and secondary windings is usually 1:1 but could be varied to give a higher current pulse,

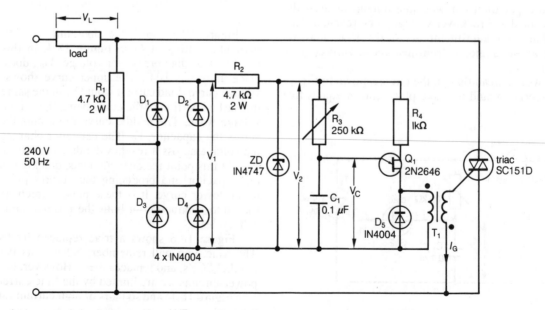

Fig. 18.7 *A triac control circuit triggered by a UJT*

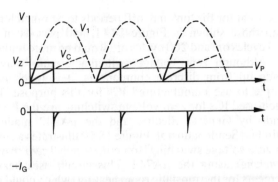

Fig. 18.8 *Waveforms for the circuit in Figure 18.7. Upper curve represents voltages and lower curve gate current*

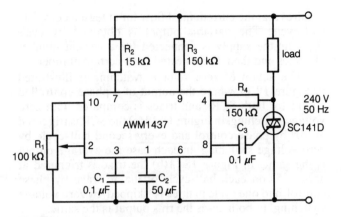

Fig. 18.9 *A triac triggered by an IC*

at a lower voltage, to the gate of the triac. The diode, D_5, across the primary of the pulse transformer functions as a flywheel diode. This tends to keep the current flowing in the one direction by reducing the emf of self induction at the instant the current pulse ceases, to a very low value (about 0.6 volt).

The period of oscillation of the UJT relaxation oscillator is such that the time between pulses is no greater than 0.01 second, the time of a half-cycle at 50 hertz. If greater than this no triggering could take place. Pulse periods less than this will cause triggering at any part of the half-cycle. It must be noted that if the pulse period is quite small, say 0.001 second, only the first pulse will be effective since after it has triggered the triac, the voltage across the triggering circuit falls to about 0.7 volt and the oscillation ceases. This fall in voltage occurs each half-cycle, so the action of the UJT oscillator is synchronised with the half-wave pulses provided by the mains through the bridge rectifier.

18.4 Triggering with an IC

As mentioned previously, integrated circuits (ICs) have been manufactured to perform many varied functions. One function includes the triggering of thyristors. An IC may be manufactured to include many circuit features such as constant control voltage, temperature compensations and control functions. One such IC has a type number AWM1437 and will deliver trigger pulses after the start of each half-cycle in proportion to the voltage across two of its terminals. There is no necessity to understand, or even know, the internal circuit of an IC and, as long as its function and circuit connections are provided by the manufacturer, it can be connected into a circuit.

In Figure 18.9 the AWM1437 IC is used to control the current to a load controlled by a triac. The position of the moving contact on the variable resistor R_1 varies the triggering angle, α, applied to the triac each half-cycle. This in turn controls the conducting angle of the triac, θ, which controls the current to the load. In common with most triggering circuits the trigger angle may be at any part of the half-cycle. The triggering pulse is applied to the gate of the triac via capacitor C_3.

18.5 Zero voltage switching

Phase control of thyristors gives excellent control of the output but does have one serious disadvantage. The very fast rise time of current at switch-on (particularly when at the middle of a half-cycle) can cause severe radio frequency interference (RFI) by generating random noise at these frequencies. This is because the sudden increase in magnetic flux accompanying the sudden increase in current excites numerous parts of the system, composed of random inductances and capacitances to resonate at their natural resonant frequencies. It is possible to use low bandpass filters (see Section 5.2) to eliminate this form of pollution and many countries have strict regulations in regard to RFI. With careful design of low-pass filters, and sometimes electrostatic and electromagnetic shielding, RFI can be eliminated from phase control circuits.

One method of elimination, without having to incorporate filters, is to switch the circuit only when the ac cycle passes through zero voltage (at 0^0 and 180^0). This is termed *zero voltage switching* (Fig. 18.10) and

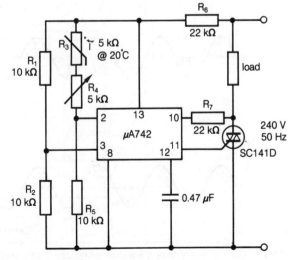

Fig. 18.10 *A zero voltage switching heat control circuit*

requires that the current must flow for at least a complete half-cycle. The variable output is obtained by *cycle skipping*: the supply is connected for a certain number of cycles and then disconnected for a certain number.

The action of zero voltage switching is illustrated in Figure 18.11 where the output of a phase controlled circuit for both SCRs and triacs is compared with zero voltage switching. In Figure 18.11(a) the SCR is triggered at 90° by phase control and every second half-cycle by zero voltage switching. In each case the average output is the same. In Figure 18.11(b) the triac is triggered at about 95° on each half-cycle of the ac wave by phase control, and one cycle in three is 'skipped' by zero voltage switching. In both cases the rms output is the same.

It must be obvious that zero voltage switching could not be used for lighting control—and even for motor control the pulsating torque produced by the 'on' and 'off' periods would be completely unacceptable. However, for heating control the method is eminently suitable since the thermal inertia of the heating elements and their surroundings can average the output. This is similar to energy controllers or simmerstats used on electric ranges.

It is usual for the 'on' and 'off' periods to be much longer than those shown in Figure 18.11, and periods of say 400 cycles on, and 200 cycles off, would be more realistic.

Although zero voltage switching circuits may be made up from discrete components, it is much more simple to use manufactured ICs for this purpose. Two often-used ICs for zero voltage switching are the PA424 made by General Electric and the μA742 made by Fairchild Semiconductor. Figure 18.10 illustrates a circuit for zero voltage switching (sometimes called *zero crossing switching*) using the μA742. This circuit uses heating elements for thermostatic room heating (which could also include a fan, which is not shown). The variable resistor (R_4), sets the desired temperature, and the NTC thermistor (R_3), through the IC, keeps the temperature constant by adjusting the on-off heating periods via the triac.

In operation, R_6 acts as a dropping resistor and R_1 and R_2 provide a reference voltage at pin 3. If the voltage at pin 2, adjusted by R_4 and the variation of R_3, rises above the voltage at pin 3, a trigger pulse is applied to the triac every time the voltage wave passes through zero—as in Figure 18.12. This is known as *on-off control*

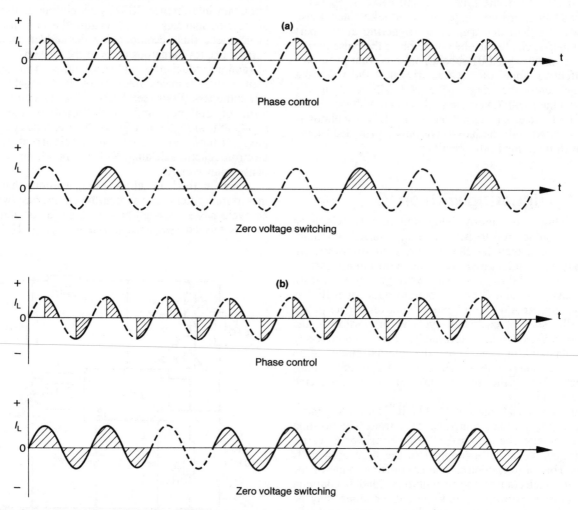

Fig. 18.11   *Comparison of current waveforms of phase control and zero voltage switching for both SCRs and triacs: (a) control by SCR, (b) control by triac*

and is similar to the simple snap action of a thermostat, but there are no moving parts and the reliability is usually much better.

Another system, called *proportional control*, can also be used. In this, as the temperature approaches the set level the number of 'on' to 'off' cycles gradually reduces until the temperature is reached when they remain off. Similarly, if the temperature falls, the 'on' periods are only minimal unless there is a *rapid* fall: when the 'on' periods become longer until temperature rises again towards the set point. The 'proportional' system usually leads to less 'overshoot' in the desired set temperature.

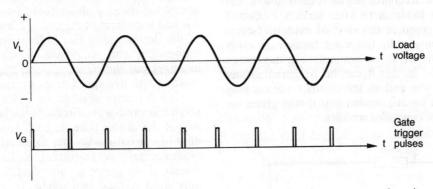

Fig. 18.12   *For zero voltage switching, gate pulses are applied to the triac gate each time the wave passes through zero*

Unit 18 **SUMMARY**

- Phase control of thyristors means allowing conduction for only a part of an ac cycle.
- The angle in a half-cycle when triggering of a thyristor takes place is called the triggering angle, represented by the symbol α.
- The conduction angle in phase control is the angle in the half-cycle in which conduction takes place, and is represented by the symbol θ.
- Phase control can use either diacs or UJTs, and the triggering angle α can be varied by adjusting the value of the RC resistor.
- Special ICs are made to use as thyristor triggering devices.
- Zero voltage switching is used to prevent RFI.

Controlled rectifiers

19.1 Controlled rectification

As the SCR is a unidirectional device it must follow that it could replace a diode in rectifier service. In fact if the SCR was triggered at the start of each half-cycle it would behave in exactly the same fashion as a PN diode. However the SCR has one very big advantage over the PN diode, in that it can be triggered at any part in the half-cycle and so the rectifier output may be controlled. It is for this reason that it was given the name SCR—silicon *controlled* rectifier.

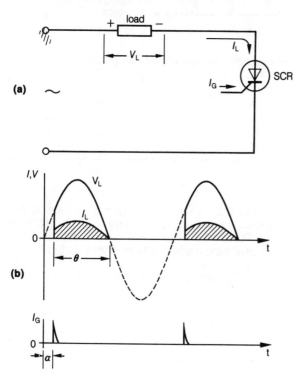

Fig. 19.1   *(a) A simplified circuit of an SCR controlled rectifier*
(b) waveforms of voltages and currents in the circuit

Figure 19.1 shows the basic circuit of a half-wave controlled rectifier. The SCR in the circuit can be triggered at any angle, α, after the commencement of the half-cycle, and will conduct for the remaining part of the cycle, θ. Although the resultant waveshape is severely distorted (from the normal half sine-wave), the current is in only one direction and hence the circuit behaves as a dc rectifier.

As in any rectifier, so long as the supply voltage is not very low, the forward voltage drop across the SCR may be neglected and we can say the full supply voltage appears across the load during conduction. By triggering at different trigger angles, the average voltage and current at the output may be varied. Filtering can provide a smoother output voltage.

Half-wave rectification is seldom considered an ideal choice but does have some applications. One is the speed control of small domestic appliances and hand-held electric drills. In these, the speed of the motor driving the appliance may be varied from a very low one to almost full speed, with almost constant torque. The rather unusual waveform of the current flowing in the armature of the driven motor has little effect on operation. The momentum and rotational velocity of the armature averages out the pulsed dc and this is reflected in the steady controlled speed of the appliance.

It is far more usual to see SCRs performing rectifier service in a full-wave circuit. In the bridge rectifier circuit two of the diodes (one on each half-cycle) perform the duty of current guides and so, in effect, only two of the diodes actually act as current blocking rectifiers. For this reason, if we desire a controlled full-wave rectifier we only need replace two diodes in a bridge with SCRs. This can be seen in the circuit in Figure 19.2.

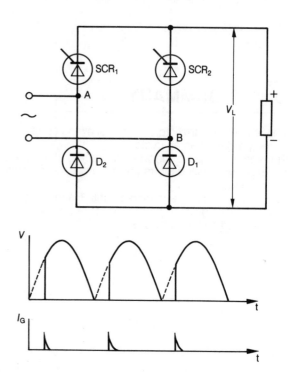

Fig. 19.2   *Circuit and waveforms for simplified full-wave controlled bridge rectifier*

The two upper devices in Figure 19.2 are SCRs, and the two lower are PN diodes. When point A is positive, current will flow through SCR_1 at a conduction angle controlled by the triggering circuit, through the load and via the guide diode (D_1) to the point B. If the SCR is not permitted to trigger, no current would flow. SCR_2 is, of course, reverse biased and cannot conduct.

When point B is positive, current flows through SCR_2, the load and guide diode D_2—as controlled by the triggering circuit—to point A. Note then that the two

SCRs shown are all that is required, with the diodes acting as current guides.

As in the bridge circuit containing only diodes, the current passes through the load in the same direction for each half-wave. However, when SCRs are used the output can be controlled from zero to full output current.

19.2 The single-phase half-wave diac/SCR controlled rectifier

Figure 19.3 illustrates a simple half-wave controlled rectifier using a diac to trigger the SCR. With the diac used, the conduction angle can be controlled from about 175^0 to zero degrees. The operation is similar to the diac/triac circuit of Figure 18.4 and all components are the same except that this circuit is a rectifying circuit and the circuit of Figure 18.4 passes alternating current in both directions.

The diac can be made to fire at any point in the half-cycle when the voltage V_C reaches about 30 volts (the breakover voltage of the diac). When this happens the voltage drop across the SCR falls to about 0.7 volt and no further current flows in the trigger circuit. The

diode, D_1, is essential in this rectifier circuit as reverse current must not flow from the gate of the SCR. This would take place in the negative half-cycle when the SCR is not conducting. Diode D_1 therefore acts as a blocking diode to prevent this from occurring.

19.3 The single-phase half-wave UJT/SCR controlled rectifier

The circuit shown in Figure 18.7, can also be used to trigger an SCR in a half-wave rectifier circuit. This can be seen in Figure 19.4, which uses current pulses generated by a UJT relaxation oscillator. The current pulses are synchronised to the mains supply and are delivered each half-cycle. The SCR only requires a triggering pulse in the positive half-cycle but, as it will only conduct when its anode is positive, the pulse received in the negative half-cycle poses no problem.

Comparing the circuit in Figure 19.4 with the one in Figure 18.7 it can be seen that the polarity of the pulse transformer has been reversed: this is so that only positive pulses of current are applied to the gate of the SCR. Current flowing out from the gate might destroy the SCR. Once again, conduction may be from 180^0 to zero degrees in each positive half-cycle.

19.4 The IC triggered SCR single-phase half-wave rectifier

Trigger pulses may be applied to the gate of an SCR by an IC in the same way it can be applied to a triac gate. The circuit for a half-wave rectifier, using an IC, can be seen in Figure 19.5. Note that this circuit is identical with the one in Figure 18.10 except the triac is replaced with an SCR.

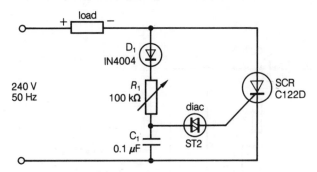

Fig. 19.3 *Circuit of a diac-triggered SCR controlled half-wave rectifier*

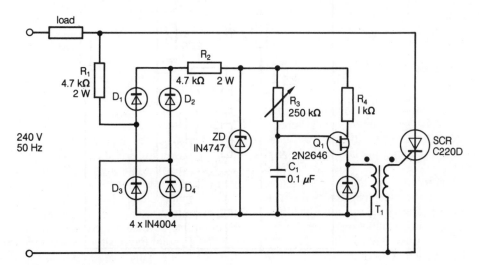

Fig. 19.4 *Circuit of a UJT-triggered SCR controlled half-wave rectifier*

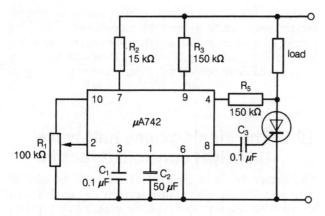

Fig. 19.5 *Circuit of an SCR-controlled half-wave rectifier triggered by an IC*

19.5 A full-wave controlled rectifier

A complete circuit of a controlled full-wave rectifier is shown in Figure 19.6. This circuit is a centre-tap controlled rectifier formed from the two SCRs, SCR_1 and SCR_2. The two SCRs are triggered each half-cycle by the triggering circuit in the lower part of the drawing but will only conduct when their respective anode is positive. This results in each SCR being triggered and conducting in each alternate half-cycle.

The supply for the triggering circuit comes from the two diodes, D_1 and D_2, which also form a centre-tap full-wave rectifier. R_1 and ZD_1 provide a regulated 16 volt supply for the UJT relaxation oscillator formed by resistors R_2-R_5, Q_1 and C_1. Adjustment of R_3 will provide output voltage from zero to 240 volts.

Adjustment of the variable resistor, R_1, in Figure 19.5, will enable conduction angles of from 180^0 to zero degrees in the half-cycle. The IC used in this circuit is quite cheap and with the addition of three resistors (apart from the variable one) and three capacitors, an extremely reliable trouble-free control results.

19.6 Rectified motor speed control

The simplest type of rectified motor control is that used in universal series motors found in such machines as drills, mixers, sewing machines and blenders. This type of control provides half-wave dc to the motor, the average value of which may be controlled from full on, to off.

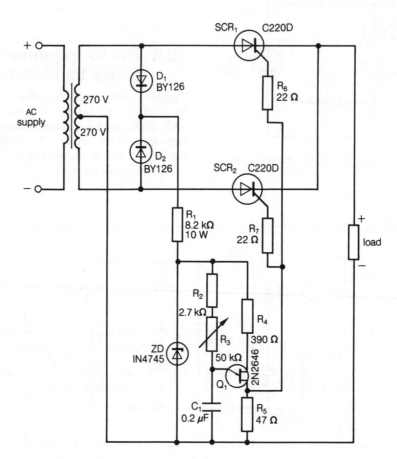

Fig. 19.6 *Full-wave centre-tap controlled rectifier circuit*

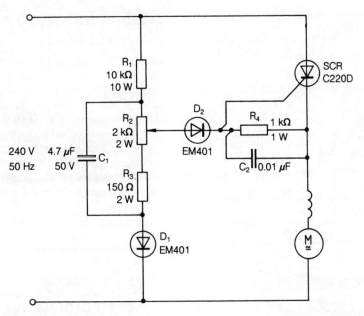

Fig. 19.7 *A universal motor speed control*

Although not specifically designed for this type of operation, normal universal motors will operate quite well in these conditions. One restriction is that for full speed operation, when using the controlled rectifier circuit motors must be designed to operate at only 67 per cent of their normal operating voltage. Another is that more frequent brush replacement may be necessary.

Figure 19.7 shows the circuit of a simple but quite effective speed control circuit for a universal motor. This circuit uses feedback control by comparing the residual back emf of the motor during the cut-off negative half-cycle, with the circuit-generated voltage across resistor R_2.

The setting of R_2 slider increases or decreases the time taken in the positive half-cycle for the rise in voltage across R_2 and C_1 to reach a value which will allow

triggering of the SCR. If the motor tends to speed up due to reduction in load, the back emf rises and diode D_2 is reverse biased until at a later time in the half cycle when the circuit generated voltage forward biases D_2 and the SCR is triggered. Conversely, if the speed tends to fall due to an increase in load, the back emf falls and D_2 conducts to trigger the SCR earlier in the half-cycle, to supply more power to the motor. This action produces a relatively constant speed for the motor at a given control setting of R_2.

The back emf is developed across R_4, and C_2 acts as a low pass filter to filter out brush 'noise' which may cause false triggering. The diode D_1 allows current to flow in only every positive half-cycle, so a cheaper polarised capacitor may be used, and heat dissipation in R_1 is reduced.

Unit 19 **SUMMARY**

- The SCR can be used as a controlled rectifier.
- In a single-phase bridge rectifier only two diodes need to be replaced by SCRs to obtain controlled rectification.
- SCRs can be triggered with diacs to provide controlled rectification.
- A UJT relaxation oscillator circuit can provide triggering pulses for an SCR controlled rectifier circuit.

- Full-wave controlled rectification can be obtained from a centre-tap rectifier with SCRs replacing the two diodes.
- One application of controlled half-wave rectification is in providing speed control to small motorised appliances.

Fig. 20.1 *Two examples of digital and analog readouts: (top) digital and analog clock, and (bottom) digital and analog multimeter*
DICK SMITH ELECTRONICS

Unit 20

Logic 1—digital concepts and logic gates

20.1 What is 'digital'?

Information signals, i.e. electrical patterns which may contain information of, say, metering, written matter or music, can be represented in two basic forms. These two forms are either *analog* or *digital*. One familiar example of this is the signals recorded on vinyl records (which are analog) and those recorded on compact discs (which are digital). Another familiar example is the digital multimeter which displays its readings in digits (single numbers), and the pointer-type or analog meter which displays its readings by the amount the pointer is deflected (Fig. 20.1).

The words analog and analogous come from the same Greek word meaning 'according to proportion'. We can say that the deflection of the pointer on the speedometer of a car is analogous to the car's speed, so the speedometer is an analog readout. Another familiar analog readout is that provided by the hands of a clock, or watch—which move continuously to represent the passing of time.

The word digital derives from the word for finger (or digit) and originally a caveman may have held up two fingers when asked how many sabre-toothed tigers he had dispatched that day. In present day terms, a digital clock or watch represents the passing of time by displaying the digits representing the hours and minutes and such. The difference between a digital and analog clock (Fig. 20.1) is that the digital clock changes in steps (one minute or one second at a time) while the analog clock changes continuously. We could sum up these two representations by saying:

- analog is equivalent to *continuous*;
- digital is equivalent to *discrete (step by step)*.

Because digital representation has a discrete nature, there is no ambiguity when interpreting the reading of a digital quantity. When interpreting the reading of an analog quantity, however, the reading is open to possible error.

Analog or digital signals may be represented by various quantities but a well recognised form is by voltages. Figure 20.2 illustrates an analog form of voltage variation and also a digital representation. The main differences may be summed up as follows:

- An analog signal voltage may vary, in time, to an *infinite number of amplitudes* and be either positive or negative in polarity.
- A digital signal voltage has only *two amplitudes, zero or a fixed voltage*—but consists of a number of pulses between these two limits.

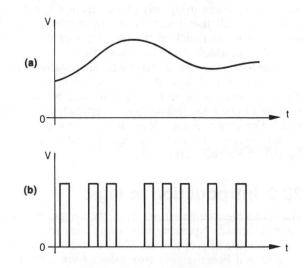

Fig. 20.2 *Example of (a) analog and (b) digital voltage signals*

A *digital system* could be a combination of various devices such as electrical, electronic, mechanical or photoelectric—which can be arranged to represent quantities digitally. In everyday life most quantities are analog and it is these quantities which are being monitored, measured and controlled. To represent them digitally they are converted to a digital quantity by an *analog-to-digital (A/D) converter*. In very many instances a digital signal is preferred as it is far less prone to distortion or degradation by transfer processes. Most compact discs have a code somewhere on the label containing three letters, each of which could be A or D. The A stands for analog and the D for digital. The three letter code indicates what processes in the production are either analog or digital. A compact disc marked 'ADD' means it was originally recorded analog, processed digital, finally recorded digital. If marked 'DDD' all processes would have been digital. On the final replay, the digital information on the disc is again converted to analog signals (by a digital-to-analog converter in the player) so that we can hear the actual recorded sound—as our ears could not interpret the digital signal.

The digital signal, as mentioned before, could be two levels of voltage—but it could also be based on two different frequencies, or on-off light action, holes or no-holes in punched tape, pits or no-pits in the surface of a compact disc, or the opening and closing of electrical contacts.

The opening and closing of electrical contacts to represent a digital signal occurs in the rotary dial of the older automatic telephones—which transmit a digital signal to the automatic exchange so that the correct called number can be selected by the exchange equipment. (Incidentally, digital telephone dialling was invented by A.B. Strowger in 1889.)

As either of *two* states can be represented in a digital system we only need two numerals to describe the system condition. In electrical terms this could be either 'on' or 'off' and in numerical terms we could say the condition was 1 or 0. When using only two digits in a numbering system we call the system *binary* (a system made up of two parts) and each of the two digits of the system we call a *bit*, which stands for *bi*nary dig*it*.

The two states of a digital system are usually referred to as *logic* **1** and *logic* **0**.* By convention, in simple electronic circuits, logic **1** is the more positive or 'high' state and logic **0** the less positive (practically zero) 'low' state. These two distinct logic states allow the use of circuits which will readily recognise, and be operated by, only these two states.

20.2 Introduction to logic

The word *logic* comes to us from the Greek, where it meant originally 'speaking of reason'. This term meant the examining of propositions which could be either *true* or *false* and because only two values were considered it was termed *bivalent* (two-value) logic. The Greeks wrote plays around the rules of logical deduction.

In 1847 George Boole developed a system of algebra useful in solving logic problems, and in 1854 put all his ideas together in a book titled *An Investigation on the Laws of Thought on which are Founded the Mathematical Theories of Logic and Probability*. The first proposition of using electrical circuits for logic arguments was proposed by Charles Peirce in 1890 but the actual development of rules to make the best use of switches, relays and circuits was prepared by the American Claude Shannon in 1937 (*A Symbolic Analysis of Relay and Switching Circuits*).

From the work of these men modern electronic logic circuits have been made possible. Not only relatively simple switching circuits, but complex machinery control, circuit switching control, calculators and computers.

20.3 Logic notation

Charles Peirce produced circuits which are still used today to explain the logic notation proposed by Boole. These two circuits are given in Figure 20.3.

In the circuit marked '**AND**' it is evident that *both* push-buttons A **AND** B must be operated in order for the lamp to light. In the circuit marked '**OR**' it is also evident that if *either* A **OR** B pushbutton is operated the lamp will light. From these two circuits we could make two statements. For the **AND** circuit we could say:

'Operation of A **AND** B will give an output at Z'.
For the **OR** circuit we could say:

'Operation of A **OR** B will give an output at Z'.
However by using Boole's notation these statements may be more suitably expressed. Boolean algebra uses the following notations:
  • means **AND**
  + means **OR**
so the above two statements may now be written:

$A \bullet B = Z$
$A + B = Z$

A further notation is illustrated in the circuit of Figure 20.4. In this circuit the operation of the push-button (which is normally closed) de-energises the lamp, which goes out.

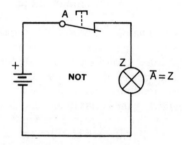

Fig. 20.4 *The **NOT** function or invert switching*

We could say from this circuit:

'**NOT** operating A will give an output at Z',
so this is termed a **NOT** function. In Boolean algebra this is written as:

$$\bar{A} = Z$$

(the bar above A meaning **NOT**).

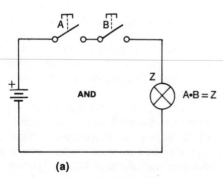

(a)

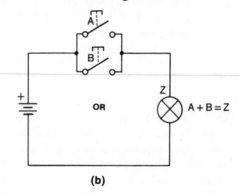

(b)

Fig. 20.3 *Concept of (a) **AND** and (b) **OR** logic gates*

---

* In this book, the two digits of a binary system **1** and **0**, are shown in bold type to avoid confusion with the digits 1 and 0 of the normal decimal counting system.

As circuits become more complicated, it is simpler to explain their function by a Boolean equation than by a written description. In Figure 20.5 is the circuit of a switching arrangement which can be described by the Boolean equation:

$$A \bullet B \bullet (C + D) = Z$$

To describe this in a sentence we would have to say:
'Operation of A **AND** B **AND** either C **OR** D will give an output at Z'.

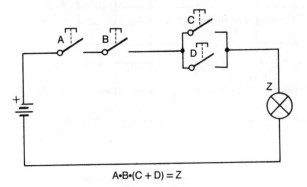

Fig. 20.5 *A combination of **AND** and **OR** functions*

20.4 Logic symbols

As will be seen later the switches in the preceding circuits may be replaced by different electronic functions but may be represented in diagrams by symbols, no matter what the actual circuitry may be. The two main symbols are for **AND** and **OR** functions. The other symbol for the **NOT** circuit may be combined with the **AND** and **OR** to form the combination **AND** and **NOT** or **OR** and **NOT**. We may also incorporate these combination functions in the one circuit and then we refer to them as **NAND** and **NOR**. These symbols are in Figure 20.6.

It is usual to refer to these circuits as *gates* as the output may be open or closed depending on the operating conditions of the inputs. In Figure 20.6 the symbols have been provided with two inputs, but a greater number (say up to eight) may be provided.

20.5 Truth tables

Still using the push-button concept, we call the *operation* of a push-button logic **1** and the *non-operation* logic **0**. In the electronic circuits we normally employ, as we have seen in Section 20.1, logic **1** can be represented by a higher voltage (within limits) and logic **0** by a voltage between zero and a lower limit. However, when first encountering logic gate operation, it is sometimes easier to visualise the input conditions when thinking of the operating of a push-button switch.

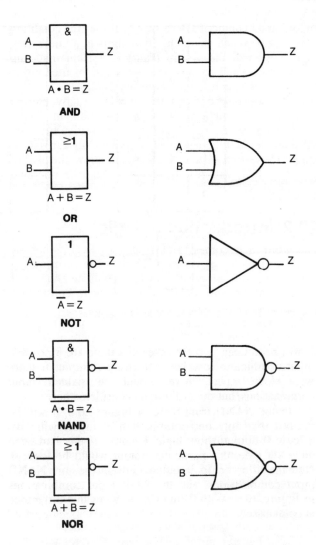

Fig. 20.6 *Logic gate symbols with their Boolean equation. The symbols on the left are Australian Standard (AS 1102, Part 9 — 1986) Binary Logical Elements and are the correct symbols to use. The symbols on the right are obsolete symbols but are included for reference as older, or some overseas, publications may use them. The older symbols are deprecated by the Australian Standards Association*

To show the possible outputs for various gate inputs we can arrange a table known as a *truth table* (see Figure 20.7). These tables give all the combinations of inputs and show the output for each combination. Figure 20.7 has the input and output conditions for three logic gates: the **AND, OR** and **NOR**. Compare the output of the **OR** and **NOR**: for the same inputs, it can be seen that all the outputs are opposite, in logic sense. It is quite simple, then, to visualise the output of a **NAND** gate, which will be opposite, in logic sense, to the **AND** gate.

When logic gates have two inputs there are four possible output conditions from the number of input combinations. Figures 20.8 and 20.9 illustrate the input combinations and possible outputs for 3-input logic gates.

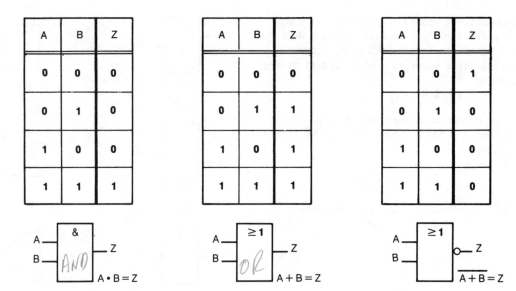

Fig. 20.7 *Truth table for three two-input logic gates*

It will be noticed in these cases that there are 8 possible input combinations which will give an output. If there were four inputs there would be sixteen input combinations, but this will not be considered here.

In the **NAND** truth table in Figure 20.8, it will be noticed that only one combination of inputs will give a logic **0** output: three logic **1** inputs. If this had been an **AND** gate, this particular output would be logic **0**. Note the difference in the output between the same **NAND** input combinations, and the **NOR** input combinations in Figure 20.9. With 3-input logic gates this difference is emphasised.

A	B	C	Z
0	0	0	1
0	0	1	0
0	1	0	0
0	1	1	0
1	0	0	0
1	0	1	0
1	1	0	0
1	1	1	0

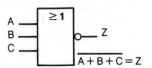

Fig. 20.9 *Three-input **NOR** gate and truth table*

20.6 Circuits of logic families

It was mentioned in Section 20.6 that the push-button function of the logic gates is actually performed by electronic functions, and we will now have a brief look at some of these. (It is not suggested the student should remember or even understand these circuits—they have been included for background information only. At times some background information makes it easier to understand the actual subject matter.)

The various logic circuits are called families, as the different logic gates may be made from each method. Some logic families are:

A	B	C	Z
0	0	0	1
0	0	1	1
0	1	0	1
0	1	1	1
1	0	0	1
1	0	1	1
1	1	0	1
1	1	1	0

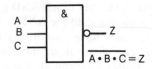

Fig. 20.8 *Three-input **NAND** gate and truth table*

DL   diode logic
RTL   resistor transistor logic
DTL   diode transistor logic
ECL   emitter coupled logic
IIL   integrated injection logic
TTL   transistor-transistor logic
MOS   metal oxide semiconductor logic
CMOS   complementary metal oxide semiconductor logic

As we stated previously, only TTL and CMOS are in widespread use today but we will briefly look at some other types for comparison. In any logic gates there is one very important fundamental condition concerning the inputs (whether it be a two input or multiple input gate): *all* the inputs must be *either* connected to a logic **1** or a logic **0** condition. No input must be left *floating*, i.e. not connected to either of these two conditions. If they are left floating, they could assume an undesired input condition and give a spurious output. In many cases a designer may be using, say, a 4-input gate when only two inputs are needed. The two unwanted inputs are not ignored, but are deliberately connected either 'high' or 'low' depending on the circuit: this must *always* be strictly adhered to. In the circuits to be briefly examined, the inputs are marked *A, B, C* and the like, and the logic condition to which all the inputs must be connected are marked **1** and **0**. In all cases this is the positive supply line for logic **1** or 'high' and the negative supply line for logic **0** or 'low'.

DL—Diode logic The simplest electronic logic circuits are the diode logic circuits of Figure 20.10. Here we have **AND** and **OR** gates each with three inputs.

Consider the DL **AND** gate. If *A, B,* or *C*, or any combination of them is connected to logic **0** (frame or negative) the diodes are forward biased, and current flows through them from the positive line and resistor *R*. The voltage at *Z*, then (with reference to frame), is practically zero (actually it is the small voltage drop across the diodes, about 0.6 volt). This is logic **0**.

Now if *A, B* and *C* are connected to the logic **1**, 5 volts positive, the diodes are reverse biased. Now, as no current flows through the resistor *R*, the voltage at *Z* is 5 volts positive, which is logic **1**. As this is an **AND** gate, all inputs must be logic **1** to give an output of logic **1**.

In the DL **OR** gate of Figure 20.10, it can be seen if either *A, B, or C* are connected to 5 volts positive, logic **1**, those diodes in the connected line are forward biased. Current flows through resistor *R* giving an output at *Z* of logic **1**, actually about 4.3 volts (5 volts less the diode voltage drop of about 0.7 volt). Note that any inputs not connected to logic **1** *must* be connected to logic **0**. If all the inputs are connected to logic **0** then, naturally, the output will be logic **0**. It can be seen, then, that if any, or all, inputs are connected to logic **1** the output must be logic **1**. These are the conditions for an **OR** gate.

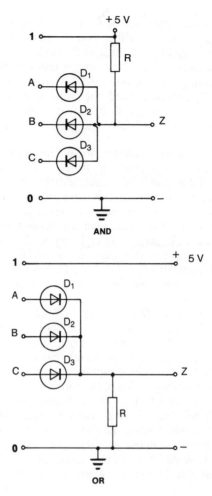

Fig. 20.10   *DL* **AND** *and* **OR** *logic gates*

DTL—diode transistor logic The DTL logic family is similar to the DL except that a BJT, used as an inverter, is added. Referring to Figure 20.11 it can be seen that the **NOR** DTL gate has the same circuit as the **OR** DL gate with the addition of two resistors and the BJT. If all inputs are made logic **0**, no base current flows in Q_1, so no collector current flows. This means that there

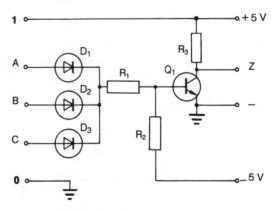

Fig. 20.11   *A DTL* **NOR** *logic gate*

is no voltage drop across R_3, so Z will have an output of logic **1**. Now if either two or all of the inputs is connected to logic **1**, then each diode in a connected line is forward biased, and current flows through R_1 and R_2 providing base current for Q_1. Collector current then flows, Q_1 saturates, and the voltage at output Z is practically zero, logic **0**. It can be seen that these are the conditions for a **NOR** logic gate.

TTL—transistor-transistor logic The TTL family of logic gates is not made up from discrete components (as are those previously discussed) but is produced as an IC by intricate manufacturing processes. This class of logic makes use of effective multi-emitter BJTs in their internal construction. Each emitter becomes an input and some TTL gates have eight inputs. The production process means that they can also be produced cheaply.

In Figure 20.12, showing a TTL **NAND** logic gate— if all inputs are connected to logic **1**, then all emitter-base junctions of Q_1 will be reverse biased. The collector-base junction of Q_1, however, will be forward biased since Q_1's collector will only be about 1.2 volts (with respect to earth), as a result of the base-emitter voltage drops of Q_2 and Q_4. This means that current will flow from the 5 volt positive supply, through resistor R_1 and the base-emitter junction of Q_1, into the base of Q_2 causing collector current to flow in Q_2. Emitter current flowing from Q_2 will bias Q_4 to a saturated state. However, Q_3 will not conduct since the diode at its emitter and the low collector voltage at Q_2's collector prevent base current flowing. With Q_4 in a saturated state the voltage at output Z will be very low (about 0.4 volt), logic **0**.

If one, or any combination of the inputs is connected to logic **0**, then Q_1's emitter-base junctions will be forward biased. Current then flows from the 5 volt supply, through R_1 and out from the emitter(s) which is connected to logic **0**. Q_1's collector current will fall to zero causing

Q_2 and Q_4 to be turned off. Now with Q_2 off the high voltage at its collector (practically the full 5 volts) will supply base current to Q_3 which will conduct to allow about 3.7 volts to appear at the output Z (5 volts less the diode and base-emitter voltage drops). This is in the logic **1** band for TTL so one or any inputs at logic **0** will give a logic **1** at the output. This then is clearly the function of a **NAND** logic gate.

CMOS logic As MOSFETs (metal oxide semiconductor field effect transistors) have not been covered in this book, no attempt will be made to explain in detail the operation of the CMOS logic gates. The CMOS family is again made by intricate processes and has the advantage over other logic families of using very little power and not requiring resistors in the construction. Once more the automated manufacture can produce relatively cheap units as compared to discrete components.

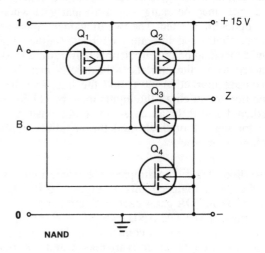

Fig. 20.13　*A CMOS* **NAND** *logic gate*

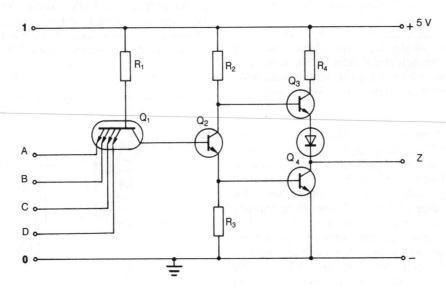

Fig. 20.12　*A TTL* **NAND** *logic gate*

The only problem with CMOS logic gates is that great care must be taken in the handling, storage and assembly of the units since any stray static voltages can permanently damage the internal circuits. The ICs containing the circuits are usually packed in a conducting foam which prevents this damage in storage and transit. During installation precautions must be taken by earthing soldering irons to the circuit under construction, keeping the IC away from plastics (which may have a static charge on them) and sometimes even earthing the wrist of the operator. Once installed in the circuit, however, there is seldom any need to worry about static damage. The internal circuit of a **NAND** CMOS logic gate appears, for interest's sake only, in Figure 20.13.

20.7 TTL logic ICs

Perhaps the most widely used logic family is the TTL logic system. Although introduced as long ago as 1964 (by Texas Instruments) it has been produced by many semiconductor manufacturers. Derivations of the original concept have kept TTL in the forefront because of its ease of use. It must be stressed here, though, that CMOS logic, because of its wider voltage range and very low power requirement (despite precautions with its handling and installation) is becoming more favoured for many installations.

The whole range of TTL products was given the name the *7400 series*, because the first product, an IC containing four two-input **NAND**s, was given this number. Many products have been added to the range but all use the first two digits of the series plus two others— for example an eight input **NAND** IC has the type number 7430 and a triple 3-input **NOR** has type number 7427. Other manufacturers, notably Fairchild, use numbers such as 9N00, 9300, 9600 and such, but usually also include the equivalent 7400 series number in their specification sheets.

Since the original 7400 series was developed, other sub-series have been made—one of these is the 7400 LS (LS standing for low-power Schottky). These place the LS between the 74 and the last two digits, e.g. 74LS27. In addition, CMOS operation at standard TTL voltages and conditions has been added, and these units use the letters HC (HC standing for high-speed CMOS), e.g. 74HC27. All are functionally similar and in almost all cases are pin-for-pin replacements.

Almost all of the 7400 series are available in one of two package types but the most widespread is the 14 pin DIL (dual-in-line)—and this is the only one we will consider here. Figure 20.14 shows a photograph of a 14 pin DIL together with the internal connections and pin connections for a 7400 IC. (Sometimes the DIL is referred to as a DIP: dual-in-line pin.)

Although usually not shown in logic circuit diagrams, the logic gates must be supplied with the correct operating voltage. In the 14 pin DIL it is almost universal for pin 14 to be the positive supply (sometimes referred to as V_{CC}) and pin 7 the negative supply (sometimes referred to as ground, or GND). Three exceptions are the 7473, 7475 and 7490. It is almost essential to refer to the manufacturer's data of pin connections before using any logic IC in a circuit.

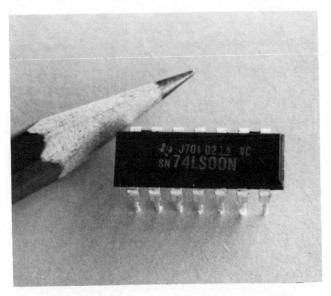
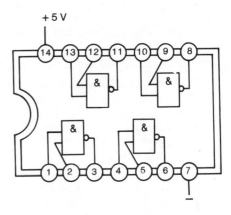

Fig. 20.14   *A 14-pin DIL package of a quad two-input* **NAND** *logic gate and a diagram of the internal connections and pin connections. This is a 7400*

Unit 20 **SUMMARY**

- Information signals may be transmitted as either analog or digital.
- Analog is continuous; digital is in discrete steps.
- Digital signals are less prone to distortion or degradation in transmission than analog signals.
- Simple logic circuits may be represented by push-button switches.
- The three basic logic gates are **AND**, **OR** and **NOT**.
- The function of logic gates can be expressed in Boolean algebra.
- The correct logic symbols to use are defined in AS1102, Part 9—1986.

- A truth table lists the output of a logic gate for all possible combinations of inputs.
- There are many logic families but the two most used are TTL and CMOS.
- The TTL family of logic gates is called the 7400 series. There are sub-series from this called 74LS00 and 74HC00, the latter being a CMOS type compatible with TTL.
- Logic circuits are usually packaged in 14-pin DIL packages with pin 7 being the negative and pin 14 the positive supply.

Logic 2—using logic, and binary counting

21.1 Power supplies for digital logic systems

To obtain the two states—i.e. 'on' and 'off' or **1** and **0**—in a binary digital system, different circuit configurations may be used. In all cases the circuits have one of two outputs which correspond to the above requirements. Over the years different 'families' of digital logic circuits have evolved and most have been largely superceded till only two remain in common use today. In Unit 20 the features of the different logic 'families' were examined but in this unit we will only be concerned with the power supply requirements for two: the CMOS and TTL types.

The CMOS logic circuits will operate at from 3 to 15 volts. Any of the standard output voltages of 5, 12 or 15 volts, as outlined in Unit 9 and obtained from fixed voltage regulators, could be used for logic CMOS operation. The output representing the binary digit **0** is zero volts (or *almost* zero volts) and the output representing the binary digit **1** is the circuit voltage (or *almost* the circuit voltage). So if the circuit voltage was 15 volts, binary digit **1** would be 15 volts and binary digit **0** would be zero volts.

To achieve these outputs the input to the circuit (in the form of binary digits **1** and **0**) must be within certain limits. An input representing binary digit **0** *must not* exceed 30 per cent of the circuit voltage and an input representing binary digit **1** *must not* be lower than 70 per cent of the circuit voltage. The region between these two input levels is called the *forbidden zone* and input voltages must not fall in this zone because an indeterminate result may occur. The input for binary digit **1** is sometimes called a *high* or *logic* **1** and the input for binary digit **0** is sometimes called a *low* or *logic* **0**. Figure 21.1 illustrates the input and output voltage limits

representing the two binary digits. The symbol V_I represents input voltage and V_O output voltage. The added *H* or *L* represents a high or low (i.e. logic **1** or logic **0**).

TTL logic circuits must operate only at 5 volts, so a 5 volt fixed regulator circuit, as outlined in Unit 9, must be used for a supply. The output representing logic **1** can be from 2.4 volts up to 5 volts (a typical value being 3.6 volts), and the output representing logic **0** can be from zero to 0.4 volt. The inputs to achieve these outputs are slightly more relaxed and a logic **0** input may be between zero and 0.8 volt, and an input representing logic **1** may be between 2 and 5 volts. The region between these limits is the forbidden zone since, as with the CMOS circuits, voltages in this region may produce indeterminate results. Figure 21.2 illustrates the input and output voltages, representing logic **1** and logic **0**, for the TTL circuits.

In both the CMOS and TTL circuits the actual voltages representing the binary (or logic) digits **1** and **0** do not matter *as long as they are in their respective defined band, or limits.*

21.2 The logic probe

When checking through logic circuits—for either method of operation, *debugging* after construction (i.e. looking for errors in assembly) or fault-finding (after failing in operation)—it is necessary to look for logic levels (i.e. voltages representing logic **1** or **0**). This may be done with a multimeter or a CRO but both methods are rather slow and sometimes ambiguous, since voltage levels may be misinterpreted. It is far better to use a *logic probe* for the purpose.

With a logic probe you can tell at a glance what the logic state of the point in the circuit under test is, by applying the probe to that point and seeing which of two LEDs are glowing. One LED indicates logic **1** and the other logic **0**. Power for the probe is obtained from the circuit under test, and is usually 5-15 volts.

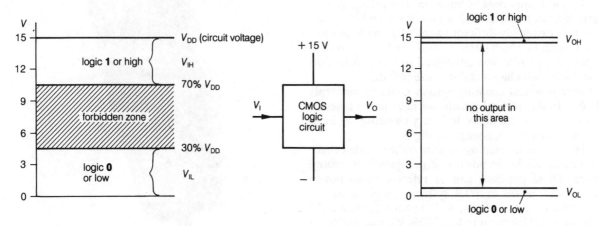

Fig. 21.1 *Input and output voltage levels of logic **1** and logic **0** for CMOS logic circuitry, when operating at 15 volts*

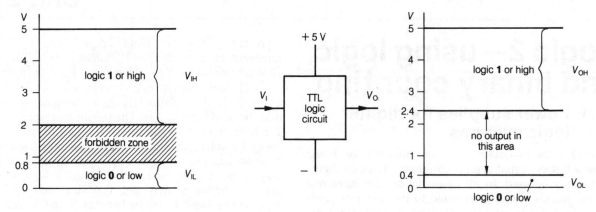

Fig. 21.2   *Input and output voltage levels of logic **1** and logic **0** for TTL logic circuits*

This is achieved by attaching two clips from the logic probe to the circuit supply. The probes are usually arranged so that they can be switched to CMOS or TTL depending on the type of circuit in which they are used. Additionally they can usually be switched from pulse to memory: in the pulse position they will indicate the rapidly varying logic state at a certain point in an easily seen manner, and in the memory position will retain the reading last obtained until cancelled. A typical logic probe is illustrated in Figure 21.3.

21.3 The binary number system

As only two figures, or bits (**1** and **0**) are considered in digital systems we use the binary counting system to represent the digital quantities. The binary counting system can represent any quantity that can be represented by our normal decimal system but will usually take a larger number of binary digits to represent the quantity.

In normal counting, the decimal system (based on ten) is used, probably because humans have ten fingers. A very early counting system based on twelve (duodecimal) had many advantages but was replaced with the decimal system—one remnant of the duodecimal system is the term 'dozen'.

The most important element in a counting system is zero. Roman numerals did not use a zero which made the system extremely difficult to calculate with and left it as a system in which numbers could only be recorded. The concept of zero was developed in about A.D. 800 by the Arabs, who had found it in use in India.

Every practical counting system contains numerals and zero. In the decimal system we have nine numerals and a zero, giving ten digits. In binary counting, we have one numeral and zero, giving two digits.

The zero, in a counting system, indicates that there are no numerals in the position it occupies in a written number. These positions can be referred to as *power positions* and, in the decimal system, are given names (from right to left) of units, tens, hundreds, thousands, etc. If we consider the number 3029, we can make up a chart of its power positions:

Fig. 21.3   *A typical logic probe*   DICK SMITH ELECTRONICS

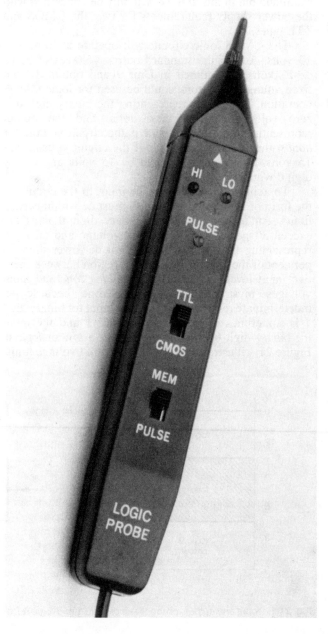

3	0	2	9
three thousands	no hundreds	two tens	nine units

In addition to the name 'power positions' we can also say the positions of the digits are *weighted* positions. The digit on the left, 3, has a greater *weight* than the other four and is called the *most significant digit (MSD)*. The digit on the right, 9, carries the least weight and is referred to as the *least significant digit (LSD)*.

In another fashion, if we use powers of ten in place of the former terms, it can be said that there are:

3	0	2	9
three 10^3s	no 10^2s	two 10^1s	nine 10^0s

(Remember $10^1 = 10$ and that $10^0 = 1$.)

In the binary system there are no names for the power positions in a number and they can only be identified by reference to decimal powers of two. The powers of two can be tabulated:

$2^0 = 1$
$2^1 = 2$
$2^2 = 4$
$2^3 = 8$
$2^4 = 16$
$2^5 = 32$
$2^6 = 64$
$2^7 = 128$ etc.

In binary numbers the power positions, from right to left, are 2^0, 2^1, 2^2, 2^3, etc. If we consider the binary number **1010**, we can make up a chart of its power positions:

1	0	1	0
one 2^3	no 2^2	one 2^1	no 2^0

The numeral **1** indicates that there is one of that power of two, in a certain power position, and **0** indicates that there is no power of two in that power position.

Binary numbers are a succession of numerals (**1**s) and zeros (**0**s), which in an electronic circuit could be the on or off state of a series of lamps or the presence or absence of a voltage, etc. However, we must be able to convert them to a decimal number, which is much more easily understood. The easiest conversion method is to allocate to each power position in the number the corresponding decimal power of two. To do this the binary number is written down with spaces between each digit. Then below each numeral (**1**) the corresponding power of two, according to its power position, is written in. The decimal numbers, which are the equivalent of the

powers of two just written in, are added to give the decimal equivalent of the binary number.

The binary number **1010** can be written down as described above and the powers of two, in their respective power position, placed below each **1** numeral. As there are no powers of two in the positions of the zeros a dash is placed in these positions.

1 0 1 0
2^3 — 2^1 —

The decimal equivalents of the powers of two, represented above, are now added to give the decimal equivalent number.

$$8 + 2 = 10$$

The decimal equivalent of binary **1010** is 10. The decimal number 10 is called 'ten' but the binary number **1010** has no name and can only be described as 'one, zero, one, zero'. Example 21.1 again shows the process of binary to decimal conversion.

Example 21.1
Convert the following binary numbers to decimal:
  (a) **111101**
  (b) **10011011**
  (c) **10001011001**

Set out the numbers and place the decimal powers of two below them.

(a)   **1 1 1 1 0 1**
2^5 2^4 2^3 2^2 — 2^0
$= 32 + 16 + 8 + 4 + 1 = 61$
(the equivalents are added)

Answer: The decimal number is 61.

(b)   **1 0 0 1 1 0 1 1**
2^7 — — 2^4 2^3 — 2^1 2^0
$= 128 + 16 + 8 + 2 + 1 = 155$

Answer: The decimal number is 155.

(c)   **1 0 0 0 1 0 1 1 0 0 1**
2^{10} — — — 2^6 — 2^4 2^3 — — 2^0
$= 1024 + 64 + 16 + 8 + 1 = 1113$

Answer: The decimal number is 1113.

Appendix E gives a list of powers of 2, to 2^{21}, and their decimal expanded number.

At times it is necessary to know the equivalent binary form of a decimal number. This could arise when it is necessary to set up a binary number as a reference, for example in a digital controlled circuit. As a binary number is made up from powers of two, it becomes necessary to break up a decimal number into powers of two to convert it to binary.

The procedure to convert a decimal number to a binary number is as follows:

  1. Divide the number by the highest power of two (in its decimal form) which will leave a remainder less than that power.

2. Divide the remainder by the next highest power of two which will leave a remainder less than that power.
3. Continue this process until there is no remainder.
4. Set out the powers of two, that were used to divide by, in their relative power position in line, placing a dash in those power positions that were not required.
5. Place a **1** below each power and a **0** below each dash and this is the binary number.

An instance of the above process is worked through in example 21.2.

Example 21.2

Convert the following decimal numbers to binary:

(a) 24
(b) 82
(c) 113

(a) The highest power of two that can divide into 24 is 16 (2^4)

so, $24 \div 16 = 1$, remainder 8.

Now the remainder of 8 is divided by the highest power of two that will give a remainder less than that power. This of course is 8 (2^3) which leaves no remainder.

$8 \div 8 = 1$, remainder zero.

Now the powers of two that were used to divide by are laid out in their respective power positions and dashes are placed where no powers of two were used.

2^4 2^3 — — —
 1 **1** **0** **0** **0**

A **1** has been placed below each power used and a **0** below each dash. This then gives us the required binary number **11000**.

(b)

$82 \div 64 = 1$, remainder 18
$18 \div 16 = 1$, remainder 2
$2 \div 2 = 1$, remainder zero.

2^6 — 2^4 — — 2^1 —
 1 **0** **1** **0** **0** **1** **0**

Answer: The binary number is **1010010**.

(c)

$113 \div 64 = 1$, remainder 49
$49 \div 32 = 1$, remainder 17
$17 \div 16 = 1$, remainder 1
$1 \div 1 = 1$, remainder zero.

2^6 2^5 2^4 — — — 2^0
 1 **1** **1** **0** **0** **0** **1**

Answer: The binary number is **1110001**.

Electronic circuits count only in binary numbers. These include digitally controlled machine tools, calculators, computers and frequency counters. In the familiar hand-held calculators, the readout is in decimal numbers but all internal counting (and calculating) is carried out in binary. This is converted to a decimal display by special internal circuitry.

Frequency meters can use digital binary counters which can count, and record, the numbers of cycles received in a certain time (controlled by an internal quartz clock). Internal circuitry then converts this to a frequency and displays the result as a decimal number.

In binary counting $1 + 0 = 1$ (and this is common to any counting system) but $1 + 1 = 10$. This is because there can be only one figure, **1**, in any power position and when another is added to it, it is shifted to the next power position. The binary number **10** (called one, zero) means that, referring to the powers of two table, we have a 2^1 and no 2^0. Therefore in decimal counting we have $2 + 0 = 2$, so it can be said that the decimal equivalent of **10** is 2.

In binary arithmetic the addition of two numbers can be made with the same rules as with decimal arithmetic except that as soon as two numerals are added their sum is a zero and another numeral, moved to the next power position left. Consider the addition to the binary numbers **11** and **11**.

$$\begin{array}{r} ^1 \\ 11\ + \\ \underline{11} \\ 110 \end{array}$$

This can be explained as: $1 + 1 = 10$, put down the **0** and carry 1. $1 + 1 = 10$ and 1 (the 1 carried) is **11**. The number now reads **110**.

Now let us analyse the above numbers in the addition. First the number **11** (one, one): this means $2^1 + 2^0$ which is decimal 3. The binary number **110** (one, one, zero) means $2^2 + 2^1$ (there is no 2^0) which is decimal 6. This then corresponds to the decimal addition, $3 + 3 = 6$.

Table 21.1 shows decimal numbers from 1 to 15 and their binary equivalents.

Table 21.1 *Decimals and binary equivalents*

Decimal			Binary
1		:	**1**
2		:	**10**
3		:	**11**
4		:	**100**
5		:	**101**
6		:	**110**
7		:	**111**
8		:	**1000**
9		:	**1001**
10		:	**1010**
11		:	**1011**
12		:	**1100**
13		:	**1101**
14		:	**1110**
15		:	**1111**

Confirm the binary numbers by adding a numeral (**1**) to each number from **1** up, in turn, to **1110**, using the rules set out previously.

Unit 21 **SUMMARY**

- In the binary system of counting there are only two digits. These are **1** and **0**.
- Each digit in a number has a weighted position.
- The highest weighted digit in a number is the most significant digit (MSD), and the lowest weighted digit is the least significant digit (LSD).
- Binary numbers may be converted to decimal, and decimal numbers may be converted to binary.

- There are two main logic systems in use: CMOS and TTL; CMOS logic circuits can operate from 3 volts to 15 volts, TTL logic circuits must operate at only 5 volts.
- Input and output voltages, representing logic **1** and **0**, must be in certain defined bands, or limits.
- Logic probes are used to determine the logic state in a circuit, i.e. whether a point is logic **1** or logic **0**.

Logic 3—universal gates

22.1 NAND and NOR for NOT

It is possible to produce any logic system from **NOR** and **NAND** logic gates. In the first instance: by joining the inputs of a **NAND** or a **NOR** gate, a **NOT** gate is created. This is because when all inputs are joined the same signal is applied to all inputs. Therefore there are only two actual conditions:

1. All inputs are logic **1**.
2. All inputs are logic **0**.

By comparing the truth tables for the **NOR** and **NAND** gates, and noting only the conditions when *all* gates are logic **1** or logic **0**, the above statement may be verified. This is shown in Figure 22.1 where we first see a **NAND** gate with the two inputs joined. This gives only one effective input, which we have called A, which supplies the gate with two identical inputs. Now if A is logic **0** the top line on the adjacent truth table shows us the output must be logic **0**. And if the input is logic **1** we see by the bottom line the output must be logic **0**. (The two centre lines of the truth table, of course, have no meaning in this case.) This is clearly the **NOT** function—the Boolean equation for the **NOT** function is placed below.

In the lower part of Figure 22.1 we see a **NAND** gate with its two inputs joined. Again we have one effective input, called *A*, feeding the **NOR** gate with two identical inputs. Now, referring to the truth table on the right, we can see that once more when both inputs are

identical (the top and bottom lines) the output is opposite to the input—or in other words the input is negated—which is the **NOT** function. Again the Boolean **NOT** equation is placed below.

It is apparent, then, that *either* a **NOR** or a **NAND** can have all its inputs joined to make a **NOT**, and Figure 22.2 shows that all these three are equivalent.

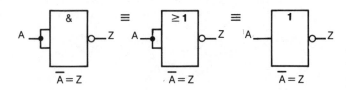

Fig. 22.2 **AND** and **NOR** gates with all inputs joined are both equivalent to a **NOT** gate

22.2 The double negation

If two **NOT** logic gates are placed in a circuit, as in Figure 22.3, it is rather obvious that the negating feature of a **NOT**, used twice, must produce an output equal to the input. This is termed *double negation*. In Boolean algebra a bar over a symbol means **NOT** (e.g. $\bar{A}$ means 'not A'). If we place *two* bars above a symbol, the second bar negates the first so that $\bar{\bar{A}}$ means 'not, not A', which of course is plain A. This can be seen in the Boolean equation accompanying the logic diagram in Figure 22.3. So, as far as logic circuitry is concerned the two **NOTs** are equivalent to a normal conductor. (Note: Appendix D of this book gives the postulates, theorems and laws of Boolean algebra, along with examples.)

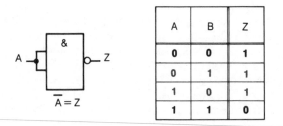

A	B	Z
0	0	1
0	1	1
1	0	1
1	1	0

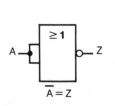

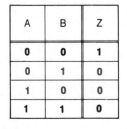

A	B	Z
0	0	1
0	1	0
1	0	0
1	1	0

Fig. 22.1 **NAND** and **NOR** logic gates can be converted to a **NOT** gate

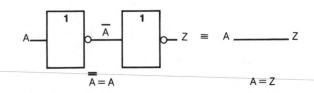

Fig. 22.3 *Double negation — two **NOTs** in series produce an output equal to an input*

As we have seen above, a **NOT** gate may be made from either a **NAND** or a **NOR** gate. It thus follows that either two **NANDs** connected as a **NOT** or two **NORs** connected as a **NOT** will also provide the effect of double negation. This is shown in Figure 22.4.

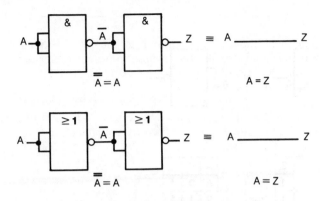

Fig. 22.4 *Double negation — two **NANDs**, or two **NORs**, connected as **NOTs** and placed in series, produce an output equal to an input*

22.3 The negated NOR and NAND

It is possible to connect two **NORs** to make an **OR** and two **NANDs** to make an **AND**. Both these conversions are seen in Figure 22.5. Note the Boolean equation that accompanies each conversion—here we have the double negation sign over the complete identity, which effectively removes the negation circle from the original gate. At this stage a student may wonder just what advantage this would have. The advantage is that only one type of gate need be used, so a stock of **NORs** can provide **NOTs**, **ORs** and the original **NORs**; likewise a stock of **NANDs** can provide **NOTs**, **ANDs** and the original **NANDs**.

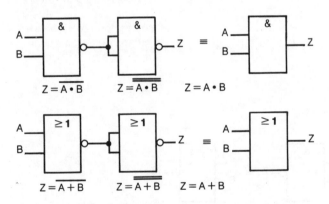

Fig. 22.5 *A negated **NAND** produces an **AND**, and a negated **NOR** produces an **OR***

feeding a signal to a third **NOR**. The input to the third **NOR** becomes $\bar{A}$ and $\bar{B}$, because both the original inputs have been inverted (or negated). Below the representation of the **NOR** gates is a truth table which sets out the logic conditions for each part of the logic circuit above. Examining the final output it can be seen that it is identical with the output of an **AND** gate. So by using three **NORs** we can produce an **AND**.

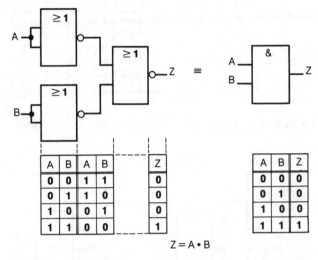

$$Z = A \cdot B$$

Fig. 22.6 *Three **NORs** connected to produce an equivalent **AND***

In Figure 22.7 a fourth **NOR** gate has been added to the logic circuit of Figure 22.6. Now the fourth **NOR** has been connected as a **NOT** (by joining the two inputs) and so its output has been inverted (or negated). Again, by referring to the truth table below, for each section of the logic circuit, we can see the output is that of a **NAND**. So by using four **NOR** logic gates we can produce a **NAND**.

Now this means that from only **NORs** we can produce any of the logic gates so far discussed. The advantage of this is that if we wish we need only stock one type of logic gate (in this case **NORs**) and from these produce any other type. Considering that logic gates are cheaply produced, and that by using two-input **NORs**, four **NORs** can be accommodated on a single IC 14-pin DIL, it seems reasonable and economic to do this. And in fact the Philips organisation produced a complete logic system using only **NOR** gates—which they called NORbits.

22.4 NORs as universal logic gates

We have seen previously that a **NOR** can be connected to become a **NOT** or **OR**. We will now see how **NORs** can be connected to become **NANDs** and **ANDs**.

In Figure 22.6 we see two **NORs** connected as **NOTs**,

22.5 NANDs as universal logic gates

By following an almost identical procedure with **NANDs** as was taken with **NORs** in Section 22.4, we can produce **ORs** and **NORs** with only **NANDs**. Referring to Figure 22.8 we see two **NANDs**, connected as **NOTs**, feeding

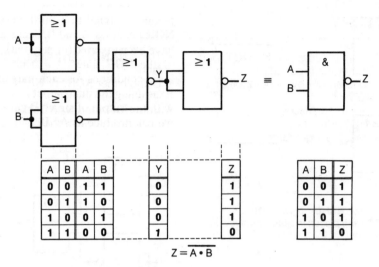

$$Z = \overline{A \cdot B}$$

Fig. 22.7 *Four **NOR**s connected to produce an equivalent **NAND***

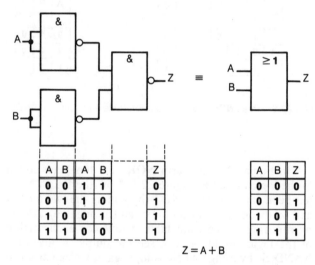

$$Z = A + B$$

Fig. 22.8 *Three **NAND**s connected to produce an equivalent **OR***

a third **NAND**. Once again the truth table is produced below, showing that the input to the third **NAND** is two negated inputs *A* and *B*. The output from the third **NAND** can be seen to be the same as the output from an **OR** gate. So with three **NAND**s we can produce an **OR**.

The addition of a fourth **NAND**, connected as a **NOT**, negates (or inverts) the output of the logic circuit of Figure 22.8, and in Figure 22.9 we can see the output is that of a **NOR**. From this it can be seen that a combination of **NAND**s can produce any logic gate so far discussed. As for the **NOR**s, it is possible to stock only one type of logic gate, **NAND**s, and from these produce any other type we wish.

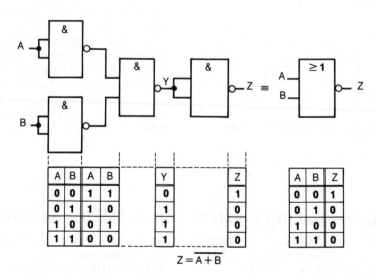

$$Z = \overline{A + B}$$

Fig. 22.9 *Four **NAND**s connected to produce an equivalent **NOR***

Unit 22 **SUMMARY**

- Both **NOR**s and **NAND**s can be converted to **NOT** logic gates by connecting together their respective inputs.
- A **NOR** can be converted to an **OR** by using another **NOR**, connected as a **NOT**, after it.
- A **NAND** can be converted to an **AND** by using another **NAND**, connected to a **NOT**, after it.
- Three **NOR**s can be connected together to form a **NAND**.

- Three **NAND**s can be connected together to form a **NOR**.
- Using either **NOR**s or **NAND**s, any other type of logic gate may be produced.
- Both **NOR**s and **NAND**s can be considered to be universal gates, as they can produce any other type of gate when connected in certain combinations.

Logic 4—further gates

23.1 The inverting buffer

Section 14.2 mentioned circuits being used as 'buffers', buffers being circuits which interface between two otherwise incompatible circuits. In this section we will examine the use of buffers between TTL outputs and loading circuits.

When TTL logic gates have an output of logic **0** they are able to accept a current of up to 16 milliamperes from the loading circuit. This is illustrated in Figure 23.1 where we can see an LED connected to the output of a TTL **NAND** gate. The 5 volt positive will allow a current of up to 16 milliamperes to flow through the LED to the gate. The LED lighting could indicate that the output was logic **0**. This is termed *current sinking* and we say that the TTL **NAND** gate, in this case, is capable of sinking this current.

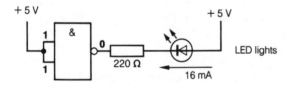

Fig. 23.1   *A TTL logic gate can sink about 16 milliamperes*

When the TTL is at logic **1**, which we sometimes call 'high', we might expect that the same current will flow from the gate output to any load attached. By the nature of TTL circuits the maximum current that can be supplied is in the order of only 400 microamperes. When the TTL gate is supplying current from its 'high' output condition we say the gate is *sourcing* the current. In Figure 23.2 we see a TTL **NAND** gate connected through a LED to the negative side of the supply, or earth, and because of the sourcing current limitations it is impossible to light the LED. We say, therefore, that the TTL **NAND** gate is unable to source the LED. In other words the TTL gate cannot operate any further device which requires a sourcing current in excess of about 400 microamperes.

Even at logic **0** the 16 milliamperes may be insufficient to operate external equipment, e.g. a relay.

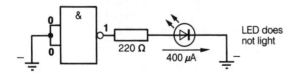

Fig. 23.2   *A TTL logic gate can source only about 400 microamperes*

In this case it is possible to insert a buffer between the logic gate output and equipment to be operated.

Figure 23.3 shows an inverting buffer connected between the output of a TTL **NAND** gate, with an output of logic **1**, and a load, in this case a relay. In this case the buffer is able to sink say 40 milliamperes, which is more than enough to operate the relay coil, from a 5 volt supply. Note the diode connected across the relay coil as a flywheel diode, to eliminate the high values of emf of self induction. In this circuit a logic output of **1** will cause an operation of the relay contacts associated with the coil.

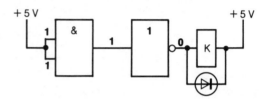

Fig. 23.3   *A buffer inverter inserted after a TTL gate can sink about 40 milliamperes when the output from the TTL gate is logic **1***

If it was required that the relay should operate when the gate output was logic **0**, this is easily arranged by inserting another inverting buffer. Figure 23.4 shows the circuit with two inverting buffers connected. The second inverting buffer negates the first so that with an output of logic **0** the second buffer will sink the current required to operate the relay. In this case the relay operation is the reverse of the circuit shown in Figure 23.3, in that it will operate for a logic gate output of logic **0**.

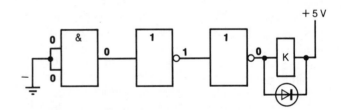

Fig. 23.4   *A second buffer inverter after a TTL gate can sink about 40 milliamperes when the output from the TTL gate is logic **0***

23.2 The Schmitt trigger

We have seen previously (Unit 20) that the voltage levels for logic **1** and logic **0** fall into defined bands. For TTL circuits logic **1** was defined as equal or greater than 2 volts, and a logic **0** equal or less than 0.8 volt. At times the incoming signal to a logic gate is accompanied by *noise*. Noise in the electrical sense means small, random instantaneous changes in voltage level. Noise may be

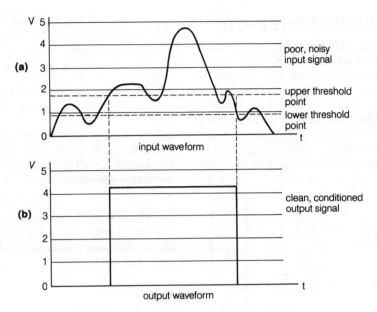

Fig. 23.5 *A poor noisy signal (a), can be 'conditioned' to a clean output signal (b), by the action of the Schmitt trigger*

produced by a poor mechanical interface at switch contacts (causing rapid changes in contact resistance) or by induced voltages in a line. Whatever the cause, the effect is that poor results may occur when the signal is degraded in this manner.

To overcome the effect of signal degradation, circuits have been developed to sense the intention of a signal voltage. Sometimes called *threshold* circuits they are usually referred to as *Schmitt triggers*. A typical Schmitt trigger will recognise a logic **1** when a rising voltage reaches a level of 1.7 volts; however, once it has recognised this as logic **1**, the input must decrease to below 0.9 volt to be recognised as logic **0**.

The effect of a Schmitt trigger can be seen in Figure 23.5. When the positive voltage first rises past the 1.7 volt threshold level, the device outputs a logic **1**. Note that the voltage then falls to below this level but the output is still logic **1**. Only when the input first decreases to below the 0.9 volt level does the output go low (logic **0**). Even a temporary rise above this does not affect the output, which still remains at logic **0**. The very poor and 'noisy' incoming signal has been 'conditioned' to a clean, neat output signal by the action of the Schmitt trigger.

The symbol for a Schmitt trigger is shown in Figure 23.6. The small figure within the symbol represents the operating characteristics of the device. This is known as the *electronic hysterisis*, or lagging behind, of the output

from the input. The output 'holds on' after the threshold has been passed until the input drops to a figure truly representing logic **0**, and 'holds on' to this level till again a rising input signal would trigger it into logic **1**.

Schmitt triggers, in practice, can be incorporated into any logic gate, but are mostly used in inverters. The Schmitt trigger inverter is placed before another logic gate to 'condition' the incoming signal so that it can be better handled. One application of this is the 'conditioning' of a signal from an analog device. This can be seen in Figure 23.7 where an NTC thermistor can give a varying signal voltage, depending on its temperature. An increase in temperature reduces the resistance of the NTC thermistor and so the voltage at the point 'X' falls. When it reaches a predetermined level (0.9 volt) the output goes to logic **1**. This gives a logic **1** for an increase in temperature (after a certain level is reached) and a logic **0** after a certain decrease. If the logic indication was required to be reversed (i.e. logic **0** for a temperature increase), a second Schmitt trigger inverter could be added after the first.

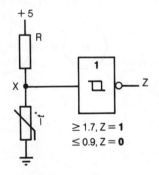

Fig. 23.7 *A Schmitt trigger inverter can produce a single digital output from an analog device*

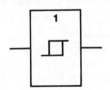

Fig. 23.6 *Symbol for a Schmitt trigger logic gate*

23.3 The EXCLUSIVE OR (EX-OR)

The **EXCLUSIVE OR**, or **EX-OR**, function is a derivation of a basic **OR** logic gate. In essence it is a system which will allow an output of logic **1** if, and *only* if, there is an operating input of logic **1** at only *one* of two inputs. In other words we can say there is an output of logic **1** *only* when the two inputs are opposite in logic level. Another way of saying this is that when the inputs are complementary, there will be an output of logic **1**. The symbol and truth table for an **EX-OR** logic gate is shown in Figure 23.8.

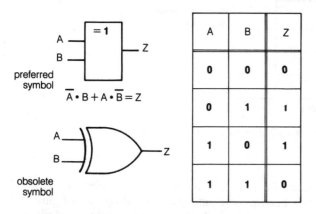

A	B	Z
0	0	0
0	1	1
1	0	1
1	1	0

preferred symbol

$$\bar{A} \cdot B + A \cdot \bar{B} = Z$$

obsolete symbol

Fig. 23.8 *Symbol, Boolean equation and truth table for an* **EXCLUSIVE OR (EX-OR)** *logic gate*

The figure also shows, lower left, the older non-standard symbol for an **EX-OR** gate, included here for reference only since it is deprecated by the Standards Association of Australia. The Boolean equation for the **EX-OR** logic gate is also shown. Note that the equation

$$\bar{A} \cdot B + A \cdot \bar{B} = Z,$$

can also be expressed as:

$$A \oplus B = Z,$$

the symbol $\oplus$ denoting the **EX-OR** function.

Adding a **NOT** gate after the **EX-OR** logic gate inverts it to an **EX-NOR** logic gate. This simply inverts

the outputs, so that the truth table will appear exactly opposite as in Figure 23.8. The Boolean equation for this arrangement is,

$$\bar{A} \cdot \bar{B} + A \cdot B = Z,$$

which can also be expressed as:

$$\overline{A \oplus B} = Z,$$

the bar simply meaning the whole function is inverted. This function, however, can also be called an **INCLUSIVE AND**, and use the shortened form of Boolean equation,

$$A \odot B = Z.$$

The **EX-NOR** logic gate could be described as being capable of producing a logic **1** output when both inputs are of the same logic level. This could be either both logic **1** or both logic **0**.

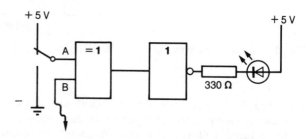

Fig. 23.9 *A simple logic probe using an* **EX-OR** *and a* **NOT** *logic gate together with a LED and current limiting resistor*

The **EX-OR** can be used as a simple logic probe if connected in a circuit as shown in Figure 23.9. The A input can be switched to either high or low (logic **1** or logic **0**). The B input is connected to the probe. Now if A is switched to high (logic **1**) the LED will light when the probe is connected to logic **1**. Then again if the A input is switched to low (logic **0**) the LED will only light when the probe is connected to a logic **0** point.

Application of the logic output of an **EX-OR** gate is when only one of any two things can be allowed to happen at any one time. In Figure 23.10 we see a diagram of a pumping system. The pump is connected to two

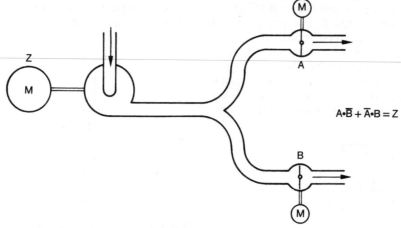

$$A \cdot \bar{B} + \bar{A} \cdot B = Z$$

Fig. 23.10 *Pumping system which can be controlled by an* **EX-OR** *logic gate*

outlets but has the capacity to supply only one. This is achieved by having a motorised valve in each of the outlets. Now if both motorised valve outlets were closed, the pump must not be able to be operated. The valves are fitted with senders so that they register logic **1** when closed and logic **0** when open. These two inputs can be connected to an **EX-OR** logic gate and the output to the pump motor control. From this we can see that only when one valve (either one) is open can the pump operate.

23.4 Logic memory

A very simple memory function using a logic system will now be examined. In this example, if a logic **1** input is made to the system, the output will go to and remain at logic **1**, even after the same input goes to logic **0**, unless another logic **1** signal on another gate cancels it. Consider the diagram in Figure 23.11.

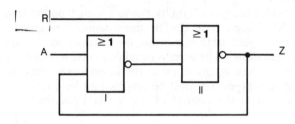

Fig. 23.11 *A simple logic memory system*

A logic **1** input to *A* produces a logic **0** at the output of the **NOR** gate I. This logic **0** applied to the **NOR** gate II produces a logic **1** at *Z*. This output is then fed back to another input of I which maintains the output of II at logic **1** even after the input reverts to logic **0** (i.e. it 'latches in'). A logic **1** input at *R* (reset) will give a logic **0** at *Z* which releases the 'latch'.

23.5 An application of logic gates

Logic systems have extensive use in control of industrial equipment. This use can be as simple as a sequence-of-operation in a machine tool or crane control, and as complex as the control of an automatic-sequence

manufacturing plant. An example of the latter is the control of a production line in a cannery. In this process such things as the supply of cans, and of the food to fill the cans, and the heating, filling, sealing, testing, labelling and packaging, must all occur in the correct sequence—each operation being dependent on another and on other factors.

If the operation of a control system is very simple, it may be cheaper and easier to use electromechanical relays; if it is extensive and complex, it will probably be carried out by programmed logic control computer. In between these two extremes, logic gate circuitry is ideal and often the obvious control method.

One application where logic gate control has wide usage is the control of traffic lights. Most traffic lights use logic control either built into the small kiosks on the roadside (for single street intersection traffic control) or in a central control centre (for extensive integrated control of a large number of road intersections).

Consider the very simple logic circuit of Figure 23.12. Here it is assumed that at a street intersection a traffic light is green, but no turn signal has been given as there is no traffic in the turning lane. If a vehicle enters the turning lane and passes over the traffic detectors (built in below the street pavement), a logic **1** is applied to the **AND** gate. At the same time the **AND** gate has a logic **1** input supplied by the direct-ahead green light which will give a logic **1** at the output and activate (through additional circuitry not shown) the 'turn' light.

It can be seen that there must be a logic **1** input from both the direct-ahead green light *and* the traffic detector before the turn light is able to be activated. When the direct-ahead green light is extinguished (and the red light activated), the turn light will also be extinguished as the output now is logic **0**. The turn light cannot be activated until once again the traffic detector *and* the straight-ahead green light produce logic **1** inputs to the **AND** logic gate.

In Figure 23.13 the traffic detector will only give a logic **1** output when a vehicle passes over it. After it has passed, the output from the detector will immediately revert to logic **0**. This in turn will produce a logic **0** at the output of the **AND** logic gate, and the turn light will be extinguished. For this reason a latching circuit is needed so that the turn light will remain on

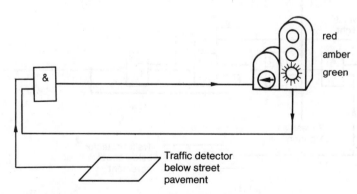

Fig. 23.12 *The traffic detector and a signal from the green light are needed to activate the turn light*

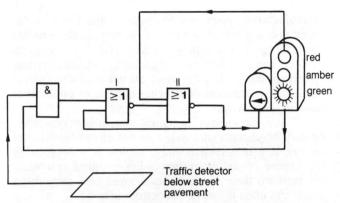

Fig. 23.13 *The turn light is latched on by a signal from the green light and the traffic detector, and only extinguished by a signal from the red light*

after the vehicle has passed the detector, enabling the turn to be made. (This is the circuit type described in Section 23.4.)

In the circuit of Figure 23.13, additional logic gates are added to the **AND** gate so that latching can take place. In this circuit, when a vehicle passes over the traffic detector a logic **1** input is applied to the **AND** gate, *together* with a logic **1** from the direct-ahead green light, which produces an output of logic **1**. This is then applied to the input of the logic memory (or latch) circuit, consisting of the two **NOR** logic gates. This will produce a logic **1** output and activate the turn light even if the output of the **AND** logic gate goes to **0** (after a vehicle has passed over the traffic detector). The turn light will remain on until the red light is activated, when a logic **1** at **NOR** logic gate II will release the latch and the turn light will be extinguished.

Another application of logic gates to traffic light control can be seen in Figure 23.14. This is a system where the lights at an intersection can be controlled by a signal from a previous set of lights down the street or by pedestrian buttons at the intersection itself. In this case a logic **1** input from a previous green light, *and* a logic **1** from a traffic detector, will produce a logic **1** output from the **AND** logic gate **1**. This output, together with the outputs of the two pedestrian buttons, is applied to the input of the three input **OR** logic gate. As it is an **OR** gate, *either* of these three inputs when at logic **1** will produce a logic **1** output from the logic **OR** gate. This output is then fed to the input of **NOR** logic gate I, which is part of the memory or latch circuit. When the latch receives a logic **1** input, even momentarily, its output goes to logic **1** and remains at logic **1** until reset. This activates the green light at the intersection.

The reset for the latching circuit is a logic **1** input to **NOR** logic gate II from **AND** logic gate 2. The logic **1** output from **AND** logic gate 2 can only be produced by a logic **1** from a timer *and* a logic **1** input from the previous red light. The action of a logic **1** output, from both the previous red light and the timer to the input of **AND** logic gate 2, produces a logic **1** output from **AND** logic gate 2. This will release the latch and extinguish the green light by giving a logic **0** output from the **NOR** logic gate II. The timer would be activated by either pedestrian button but the circuitry for this has not been shown. The timer is installed to give a reasonable time for pedestrians to cross the intersection when the lights are green. (It must be emphasised here that only a portion of the control of the traffic lights has been shown.)

These three traffic light circuits are, by necessity, very much simplified, and actual traffic light control circuits can be extremely complex. The three illustrations, however, will give the reader a basic understanding of how logic gates may be utilised to produce a sequence of events dependent on various factors considered either separately (**OR** gates) or together (**AND** gates).

Control for traffic lights can be central or local. In Figure 23.15 the set of traffic lights is controlled locally at the kiosk on the right of the photograph. The position of the detectors in the road pavement can be seen, and on the right a pedestrian is activating a pedestrian button.

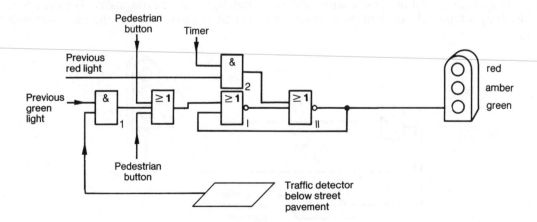

Fig. 23.14 *Green traffic light activated by either of two pedestrian buttons or a signal from a previous green light and a traffic detector*

Fig. 23.15 *Traffic lights, with roadside kiosk on extreme right, and sensors in road at bottom of photo*

Unit 23 **SUMMARY**

- A buffer is a circuit that can be placed between two otherwise incompatible circuits.
- A TTL logic gate can sink up to 16 milliamperes but cannot source any more than 400 microamperes.
- An inverting buffer can be used between a TTL logic gate output and another circuit and can sink up to 40 milliamperes when the output of the TTL gate is logic **1**.
- Adding a second inverting buffer will allow the buffer to sink up to 40 milliamperes when the TTL logic gate output is logic **0**.
- A Schmitt trigger is a circuit that will allow a logic gate to 'latch on' when a voltage (of say 1.7 volts) on an input signal is reached, and stay latched on until the input voltage falls below, say, 0.9 volt.

- A Schmitt trigger can 'condition' a poor or noisy signal.
- The **EXCLUSIVE OR** (or **EX-OR**) logic gate will only give an output of logic **1** when the two inputs are of opposite logic, i.e. **1** and **0**, or **0** and **1**.
- The addition of an inverter (**NOT** logic gate) after an **EX-OR** logic gate converts the circuit to an **EX-NOR**. This gives an output of logic **1** only when the two inputs are the same, i.e. either **1** and **1**, or **0** and **0**.
- An **EX-OR** can also be called an **INCLUSIVE AND** logic gate.
- Logic gates may be used for many processes where either one or more things, and two or more things, must occur before a further event may occur.

Logic 5—bistable devices and the JK flip-flop

24.1 Bistable circuits and devices

Most of the logic systems we have examined so far have been *combinational circuits*. These are circuits where the logic levels of the outputs, at any instant of time, have been dependent on the logic levels at the inputs. Any previous logic levels that may have been present at the input or output have no effect on the present outputs. The one exception has been the memory, or latching, circuit described in Section 23.4. This circuit is similar to the simple electrical contactor, where the contactor 'holds in' after the start button is pressed—pressing the stop button will release the contactor.

Let us consider Figure 24.1. At the top we see the conventional contactor circuit with a normally open button called *set*, and a normally closed button called *reset*. Pressing the set button will energise the coil $\frac{K}{3}$, and contact K1 will close. Taking a finger off the set button will have no effect, as the contactor has latched in. Pressing the reset button will de-energise the contactor and it will open.

Now contacts K2 and K3 are also parts of the contactor. These contacts control the voltage at the two terminals marked Q and $\bar{Q}$.

Assuming power has just been applied to the circuit, the voltage at terminal Q will be zero (which we could call logic **0**) and the voltage at terminal $\bar{Q}$ will be 5 volts (which we could call logic **1**). When the set button

is pressed and coil $\frac{K}{3}$ is energised contacts K2 and K3 change state, and the outputs at Q and $\bar{Q}$ are reversed. Q terminal now has 5 volts present and $\bar{Q}$ terminal is at zero. We could also say their logic states have been reversed. Pressing the reset button once more changes the state of the outputs Q and $\bar{Q}$.

This circuit is known as a *bistable* circuit, as it may be in either of two (bi) states, depending on what signals are made at the input. Let us now move on to electronic circuits, which will operate in much the same fashion as the contactor circuit.

24.2 The RS flip-flop

The bistable circuit is also known as a flip-flop (usually abbreviated to FF), since its output flips over from one state to another. The two outputs of an FF, called Q and $\bar{Q}$, are always complementary, i.e. if Q is logic **1**, $\bar{Q}$ is logic **0** and vice-versa. They can never both have the same logic state at the one time. The term $\bar{Q}$ simply means **NOT** Q, so that $\bar{Q}$ is always opposite in logic state to Q. The simplest and basic FF is called an RS FF, the *R* standing for reset and the *S* for set. The symbol for an RS FF is shown in Figure 24.2. There can be two conditions at each of the R and S inputs (as in any binary logic circuit): logic **1** and logic **0**.

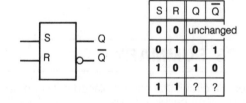

S	R	Q	$\bar{Q}$
0	0	unchanged	
0	1	0	1
1	0	1	0
1	1	?	?

Fig. 24.2 *Symbol and truth table for an RS flip-flop*

Reference to the truth table in Figure 24.2 shows that a logic **0** input at both R and S will not affect the output. A logic **0** at S and a logic **1** at R will make Q logic **0**, and $\bar{Q}$ logic **1**. If they were already in that state there would be no change. If S is logic **1** and R is logic **0**, Q will become logic **1** and $\bar{Q}$ will become logic **0**, and if they were already in that state there would be no change. Now if both R and S are at logic **1** we say the logic condition of the outputs is *unspecified* as we do not know just what the outcome could be. Because of this an RS FF never has a logic **1** at its two inputs.

Normally an FF is triggered with a pulse rather than a steady state logic condition. In the RS FF the two inputs can be in a steady state of logic **0**, this being the normal condition. A pulse at logic **1**, applied to each of the inputs R and S alternately, will cause the two outputs to alternately change from logic **1** to logic **0**. It is usual to give the output of an FF in the state at which the

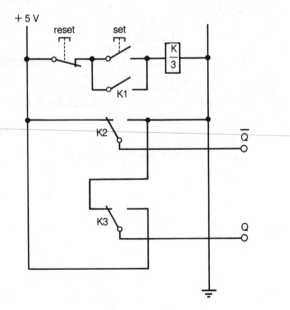

Fig. 24.1 *An electromechanical representation of a bistable circuit*

Q output is, since it is understood that the $\bar{Q}$ will be opposite. Considering this we may make a summary of the operation of the RS FF.

1. When both inputs are logic **0** (normal) the output remains what it was previously.
2. When S input is momentarily made logic **1**, the output becomes logic **1**.
3. When input R is momentarily made logic **1**, the output becomes logic **0**.
4. The condition when both S and R are made logic **1** simultaneously is never considered—because the output is unpredictable.

It must be obvious that the RS FF can *remember* the last input it has received. In this fashion it may be considered to be an elementary memory cell. This memory is limited to 'knowing' whether a logic **1** or logic **0** input was last received. Remembering that binary arithmetic counts only in digits **1** and **0**, it can be seen how the binary system is essential for electronic memory circuits.

Figure 24.3 shows an RS FF assembled using **NOR** logic gates. It is not expected that the student should memorise this circuit and it is included here for interest only. Usually FFs are purchased in ICs, as are all logic elements.

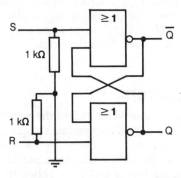

Fig. 24.3 *An RS flip-flop circuit made up from **NOR** logic gates*

24.3 Clocked flip-flops

The simple RS FF finds little application in modern circuits since it is prone to spurious results if an unwanted pulse appears at either the R or S inputs. One way of eliminating this problem is to allow the input circuit to be active only during a specific time. This time can be set with a series of pulses generated by a *clock circuit*. If the logic **1** pulses at the S or R inputs are *synchronised* (made to arrive at exactly the same time) as pulses at another input (called a *clock input* and usually abbreviated to CLK or C) they will be recognised by the circuit. If a pulse appears at either R and S that is not synchronised with the clock, it is ignored by the circuit. The *clocked RS FF*, as this circuit is called, is referred to as a *synchronous flip-flop*, whereas the RS FF discussed in Section 24.2 is an *asynchronous flip-flop*.

The symbol of a clocked RS FF appears in Figure 24.4, together with a circuit diagram using **AND** and **NOR** gates. Again, this circuit diagram is included for information only and need not be memorised by students. The truth table for a clocked RS FF is identical to the RS FF but it must be remembered that outputs will not change unless the input pulses at either the R or S terminals are synchronised with the input pulses at the clock terminal.

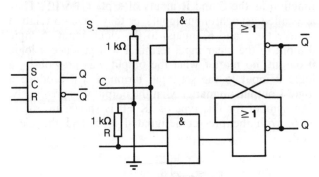

Fig. 24.4 *A symbol for a clocked RS flip-flop and a circuit representation using **AND** and **NOR** logic gates*

24.4 The JK flip-flop

Further development from the RS and clocked RS FFs led to the development of the D FF (delay or data flip-flop) and the T FF (toggle flip-flop). However all the improvements and features these other FFs possess have been combined into the standard FF now in use—the JK FF. (The letters J and K have no specific meaning and were simply chosen as they are different from all the other letters previously used and are adjacent letters in the alphabet.)

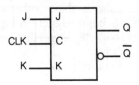

Fig. 24.5 *The symbol for a JK flip-flop*

The symbol for a JK FF appears in Figure 24.5. The internal circuit actually contains two flip-flop circuits, but that need not concern us here. The JK FF is possibly the closest approach to the ideal FF. It will operate as a clocked RS FF if we consider the J input equivalent to the S input and the K input equivalent to the R input.

If both the J and K inputs are held at logic **0** there will be no change in the output when a clock pulse is received. When the J input is held at logic **1** and the K input at logic **0**, the output will go to logic **1** and stay there when a clock pulse is received.

When the K input is held at logic **1** and the J input at logic **0**, the output will return to logic **0** and stay there when a clock pulse is received.

The JK FF will also operate as a T FF. This is done by keeping both J and K inputs held at logic **1**. In this condition the output will change alternately from logic **1** to logic **0** each time a clock pulse is received.

The JK FF can also be fitted with one or two auxiliary inputs. These are *pre-set* and *clear* and have the same function as the S and R inputs of a plain RS FF. They are called asynchronous inputs, as they have no relation (unlike the J and K inputs) to the clock input. A logic **0** input to the clear input terminal will produce a logic **0** output, no matter what the output was previously; a logic **0** input to the set input terminal will produce a logic **1** output, no matter what the output was previously. The symbol for this type of JK FF is shown in Figure 24.6, with the pre-set designated as S and the clear as R.

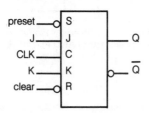

Fig. 24.6   *A JK flip-flop with preset and clear asynchronous inputs*

So far in the description of FFs we have dealt with inputs by pulses arriving at the various input terminals. The major disadvantage with pulse triggering is that the J and K inputs must be held constant (at their respective logic level) during the duration of the clock pulse. This can be eliminated by using *edge-triggered* JK FFs. Edge triggering means that the outputs can be affected by either the *leading edge* or *trailing edge* of the clock pulse. This often results in more positive triggering, but it is essential the edge of the pulse is sharp, otherwise it may have to pass through a Schmitt trigger. In Figure 24.7 we see the symbols for both leading edge and trailing edge JK FFs.

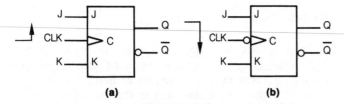

Fig. 24.7   *(a) Leading edge triggered, and (b) trailing edge triggered JK flip-flops*

The truth table for a trailing edge JK FF is shown in Figure 24.8. On the left are the conditions for the J and K inputs at the moment the trailing edge of the clock pulse arrives. In the clock column the trailing edge

J	K	clock	before		after		effect on output
			Q	$\overline{Q}$	Q	$\overline{Q}$	
0	0	↘	0	1	0	1	no change
			1	0	1	0	
1	0	↘	0	1	1	0	sets or remains set
			1	0	1	0	
0	1	↘	1	0	0	1	resets or remains reset
			0	1	0	1	
1	1	↘	0	1	1	0	toggles
			1	0	0	1	

Fig. 24.8   *Truth table for a trailing edge JK flip-flop*

of the pulse is shown, and the next two columns show the condition of the outputs before and after the clock trailing edge pulse arrives. It can be seen from the table that a logic **0** on both the J and K input prevents any change occurring in the output, no matter just what logic state the output may be. A logic **1** on the J input, and logic **0** on the K input, will set the output to logic **1**. (Remember when we say the output is logic **1**, we mean the Q output—the $\overline{Q}$ output is always the opposite to Q.)

With a logic **0** on the J input and a logic **1** on the K input, the output will reset to logic **0**. In the bottom line we have the toggle effect, which means that with logic **1** on both the J and K inputs the output will alternately toggle between logic **1** and logic **0** with each trailing edge of a pulse from the clock.

24.5 Counters

Counters are circuits which record the number of pulses received. Many industrial processes require the pulses from some type of transducer be counted and this information stored: this could be a number of cycles, of pills dropping into a bottle, or pulses from a digital transducer (a device which converts a signal from one form to another).

JK FFs make excellent counting circuits. Figure 24.9 shows a very simple type. Here we use two JK FFs which will count the number of pulses received at the clock input. Both the JK FFs receive a pulse at the same time. FF1 has its J and K input at logic **1** so it will toggle as each pulse is received. The J and K inputs of FF2 are supplied from the Q output of FF1. This means that the output of FF2 will only toggle when the output of FF1 is logic **1**. This circuit is capable of counting to **11**, which is decimal 3, with FF1 supplying the least significant bit and FF2 the most significant bit.

Let us go through the counting sequence. Assuming both outputs are logic **0**, the first pulse will toggle FF1 output to logic **1**. At the time FF2 received the trailing edge, the output of FF1 was logic **0**, so there was no change in the output of FF2. Now the next pulse received will toggle FF1 to logic **0** but as the J and K inputs of FF2 are at logic **1** (supplied by the Q output of FF1) it will toggle and have an output of logic **1**.

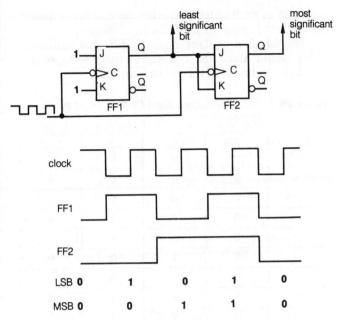

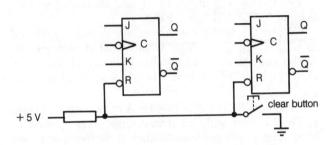

Fig. 24.9 *A simple two-bit counter with clock and output time graph, and output logic state*

If the very simple two-bit counter in Figure 24.9 was stopped at any point in its counting sequence, there would be no way the outputs could be cleared without rearranging the circuit connections or disconnecting the supply. By using a JK FF with a 'clear' input, this process can be very easily carried out. In Figure 24.10 two JK FFs with clear inputs have been connected to a high input through a resistor. This holds the clear inputs at logic **1** and they do not affect the operation of the JK FFs. When the clear push-button is pressed, and the clear inputs are brought to logic **0**, the outputs of both JK FFs revert to logic **0**.

Fig. 24.10 *A clear output push-button added to a simple two-bit counter*

The next pulse will again toggle FF1 output to logic **1** but will not affect FF2 output (because the J and K inputs were at logic **0**). The fourth pulse received will toggle FF1 output to logic **0** and, because the J and K inputs of FF2 were at logic **1** when they received the fourth pulse, FF2 output will toggle to logic **0**. This completes the sequence, and further pulses will go through it as before. The clock pulses and the output logic states of FF1 and FF2 are shown in Figure 24.9.

A slightly more elaborate counter, with LED indication, is illustrated in Figure 24.11. This is a 4-bit counter, so it is capable of counting to **1111**, decimal 15. The outputs of the FFs are taken to inverting buffers, which can sink the current through the LEDs from the 5 volt supply. The outputs could be taken to any further circuitry if required. The 'clear' input to each JK FF is

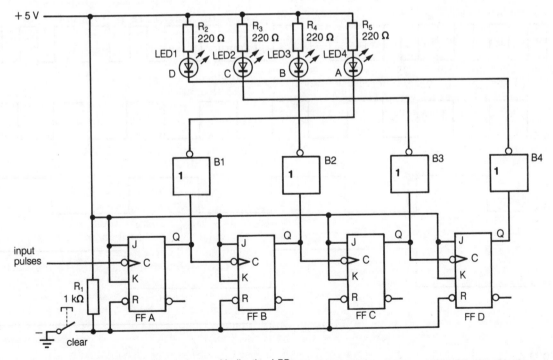

Fig. 24.11 *A four-bit binary counter with output clear and indicating LEDs*

held high through resistor R_1 and all can be made to reset to logic **0** by pressing the 'clear' push button. The LEDs have been arranged on the diagram so that LED_1 (*D*) represents the most significant bit and LED_4 (*A*) the least significant bit of any binary number displayed. This is, of course, the reverse order to the JK FFs in the diagram.

Referring to Figure 24.11, we can outline the operation of the counter. First, consider that the clear button has been pressed and all Q outputs are logic **0**. This makes the outputs of all the buffer inverters logic **1**, so the LEDs will not light.

Next, consider that a pulse arrives at the clock input of FF *A*. This will toggle FF *A*, and its output will go to logic **1**. This will light LED *A*.

When the next pulse arrives, FF *A* will toggle and its output will go to logic **0** and LED *A* will go out. At the same time, as a pulse has been completed at the clock input of FF *B*, it will toggle and its output will go to logic **1** and LED *B* will light.

The next pulse will toggle FF *A* to an output of logic **1**, FF *B* output will still remain logic **1** as a pulse at its input has not been completed. A further pulse will toggle FF *A* to logic **0**, FF *B* to logic **0** and FF *C* will output as logic **1**.

Now four pulses have been received and the LEDs will now read 'off, on, off, off,' from the left. This corresponds to binary **100** which is, of course, decimal 4.

From this point on it is better to refer to the pulse-time graph of Figure 24.12, which sets out the condition of each FF output up to 16 pulses. It can be seen that the 4-bit counter can count up to binary **1111** (decimal 15). A further pulse after this will return it to zero. The condition of each output LED for the number of pulses received is recorded in Table 24.1.

The addition of three more JK FFs and indicating LEDs would allow the counter to indicate to binary **1111111** (decimal 127) if this was desired.

Table 24.1 *The LED output state for the circuit of Figure 24.11*

Input pulses	LED D	LED C	LED B	LED A
0	0	0	0	0
1	0	0	0	1
2	0	0	1	0
3	0	0	1	1
4	0	1	0	0
5	0	1	0	1
6	0	1	1	0
7	0	1	1	1
8	1	0	0	0
9	1	0	0	1
10	1	0	1	0
11	1	0	1	1
12	1	1	0	0
13	1	1	0	1
14	1	1	1	0
15	1	1	1	1

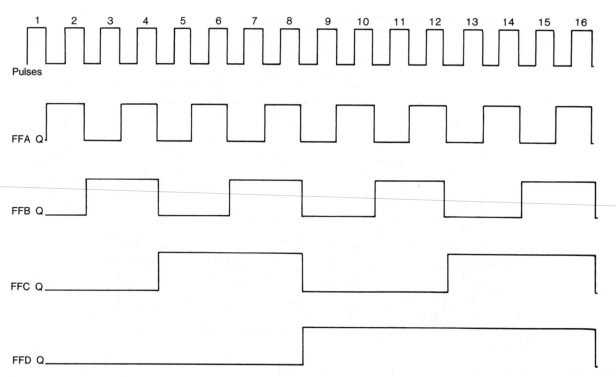

Fig. 24.12 *Pulse time graph of a four-bit counter*

Unit 24 **SUMMARY**

- The output of a bistable logic circuit may remain at one of two states after an input has been removed.
- The RS FF has a reset and set input terminal, the output depending on what was the last combination of inputs. If both inputs are logic **1** the result is indeterminate.
- To avoid spurious changes in output by unwanted pulses appearing at an input, FFs can have a clock input. These will only change outputs when both the clock and inputs act together. These are called synchronous FFs.
- The most universal FF is the JK FF.
- The JK FF can act as a D FF, T FF and clocked RS FF.

- A JK FF can have asynchronous preset and clear inputs.
- A JK FF can be pulse triggered or edge triggered.
- Edge triggering of an FF can be leading or trailing edge triggering.
- JK FFs used as toggle FFs can be connected as counting circuits.
- The clear input of a JK FF can be used to clear a counter and return it to zero.
- When JK FFs are used in counters, one FF is required for each binary digit (bit) to be counted.

Timers 1—the 555 timer

25.1 555 timer configuration

Timers used in electrical and electronic circuits can perform a multitude of different tasks. These can be for long-term timing of industrial processes (say batch mixing, where the timing period may be some hours) or for short-term timing (say a few seconds in sequence-control of manufacturing plants or in steps in motor starters). They can also be used to provide a train of pulses in a clock circuit, as we have seen in Unit 24 for logic flip-flop circuits. So it can be said a timer is any device which can be set to allow a process to continue for any set period, or to provide continuous timed pulses.

One timer which has found universal use is an IC termed a 555 timer—the number may be prefixed by other letters or numerals, depending on the manufacturer. It is made in three different package types but we will only consider the most common—the 8-pin mini dual-in-line package usually referred to as an 8-pin DIL. An illustration appears in Figure 25.1 together with an outline of the IC, showing the pin designations. (The outline is twice full size.)

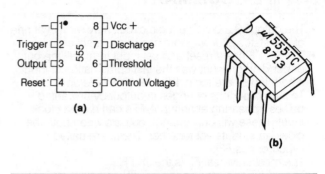

Fig. 25.1 *(a) Pin designations and (b) illustration of a 555 timer IC*

The internal circuitry of any IC is controlled by the manufacturer and the user is not able to amend or alter it. For this reason it is only of passing interest to the end-user. For interest's sake only, the internal circuitry of a 555 timer is shown in Figure 25.2.

Probably a much better understanding of the operation of a circuit can be gained from a block diagram. In Units 22 to 24, we saw block symbols used for digital logic circuits, and we can use the same scheme to illustrate

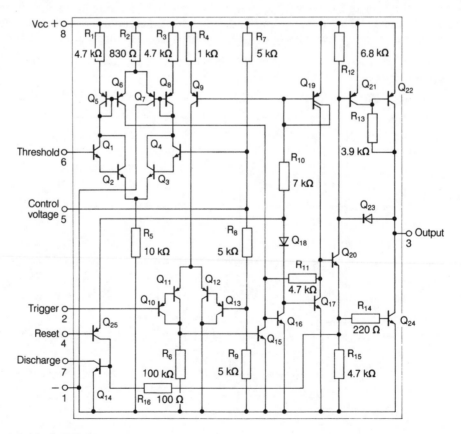

Fig. 25.2 *The internal circuit of a 555 timer*

the characteristics and operation of the 555 timer. Figure 25.3 shows that it basically consists of six parts. On the left is a two-tap voltage divider, then two op. amp. comparators, an RS FF, a discharge BJT below and an output inverting buffer.

The voltage divider connects between the positive and negative supply lines. The manufacturers state that the device can operate at supply voltages between 4.5 and 16 volts. This means that it could operate at TTL voltage (5 volts) or the 12 volt and 15 volt standard often used for CMOS logic circuits. However, unless governed by other circuit power supply requirements, the most common operating voltage for the 555 timer is 12 volts.

The power-rating for heat dissipation within the 555 timer is only 600 milliwatts, so no external heat sinking is required for its normal operation. The manufacturers, however, state that the maximum operating temperature is 70°C, and this should never be exceeded. They also state that it should not be used below 0°C, as its normal operation could be impaired.

25.2 555 timer internal operation

One of the features of the construction of the 555 timer is the internal voltage divider. This part of the internal circuit is set out in Figure 25.4, and can be seen in the centre of the circuit in Figure 25.2. It effectively consists of three resistors, each with a resistance of 5 kilohms, connected across the supply. The two tappings are connected to the internal circuitry and in addition the

upper one has an external terminal called *control*. It is quite apparent that the voltage divider tappings are at $\frac{2}{3}$ and $\frac{1}{3}$ of the supply voltage with respect to earth (negative). It is these proportions which are used in reference in the timer's operation.

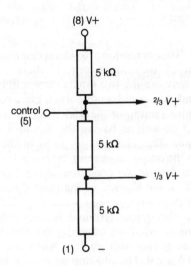

Fig. 25.4 *The internal voltage divider of a 555 timer IC*

The second feature is the two voltage comparators, Figure 25.5. A voltage comparator is basically an op. amp. The op. amp. has a very high voltage gain (A_v) and a differential input, so that if the voltage at each input (non-inverter and inverter) are *equal* there will be

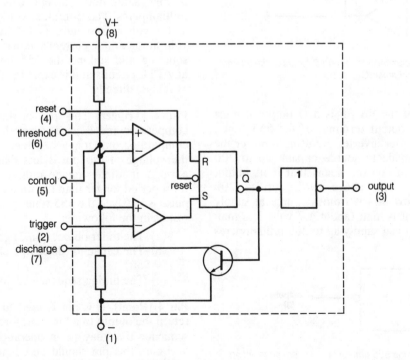

Fig. 25.3 *A block diagram of the internal workings of a 555 timer*

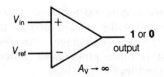

Fig. 25.5 *A comparator is a high-gain amplifier with a differential input*

no output. This condition would represent a logic **0** at the output of the comparator. If there is a voltage *difference* between the two inputs there will be an output voltage, and this would represent a logic **1** condition.

The third section of the 555 timer is effectively an RS FF, but as well as having the R and S input it has a reset input. This connects to pin 4 of the 8-pin DIL and allows the output to be reset to zero volts. This action takes place despite the condition of the flip-flop or the inputs to it. In the absence of the reset signal the output of the flip-flop will go high when R is logic **1** and S is logic **0**. The output terminal is a $\bar{Q}$ output so it is the opposite to what we normally call the output of an RS FF. The output will go low when S is at logic **1** and R is at logic **0**. The flip-flop section of the 555 timer is shown in Figure 25.6.

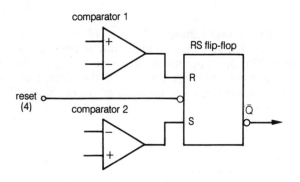

Fig. 25.6 *Two comparators feed the R and S inputs of the internal RS FF of a 555 timer IC*

As the output of the RS FF is a $\bar{Q}$ output it must be inverted for the output terminal of the 555 timer. For this we use a buffer inverter. By using a buffer the 555 timer can be made to source or sink up to 200 milliamperes. This, of course, means that if its output is low (about 0.25 volt above earth), up to 200 milliamperes can flow into it from a source at supply voltage. If the output is high (about 1.7 volts less than the supply voltage) it can supply up to 200 milliamperes

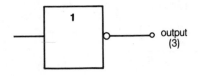

Fig. 25.7 *The output of the 555 timer IC is fed from an inverting buffer*

to a load connected between it and zero volts (negative of the supply). The inverting buffer is shown in Figure 25.7, connected to the output terminal, pin 3 of the 8-pin DIL.

25.3 Timer pin functions

To more easily understand the 555 timer, we shall look at the functional characteristics of each pin of the 8-pin DIL package. The order in which we will look at each pin will be power supply, output, input trigger and control pins.

Pin 8 (Positive) This pin, variously called V+, + supply and V_{CC} by different manufacturers, is the positive supply terminal for the 555 timer. The range of operating voltages is from +4.5 volts to +16 volts. There is no difference in the timing abilities of the timer over this operating range, and the sensitivity to timing intervals is only 0.1 per cent per volt. However, it must operate at higher voltages to be able to source its maximum current of 200 milliamperes.

Pin 1 (Negative) This is the common, earth or ground pin of the device and is connected to the most negative supply potential.

Pin 3 (Output) As a source, when high, the supply to this pin is from an internal Darlington pair and it can supply up to 200 milliamperes to a load. This capability falls to 100 milliamperes at a supply voltage of +5 volts. As a current sink, when low, another internal BJT with a low saturation voltage can accept up to 200 milliamperes. The saturation voltage varies from as low as 0.1 volt at low currents and low voltages, up to 2.5 volts at maximum current and maximum voltage. In both sourcing and sinking the 555 timer will operate with any TTL circuits and is capable of driving devices (such as relays) directly.

Pin 2 (Trigger) This pin is the input to the lower comparator and is used to set the condition when the comparator output goes to logic **1**. This occurs when the voltage at this pin is less than $\frac{1}{3}$ of V+. As this is a trigger input the voltage at this point should not remain for a period longer than the timing cycle. After a trigger pulse is received the 555 timer goes into operation. This results in the following:

(a) The timer output (pin 3) goes high;
(b) The capacitor discharge circuit (pin 7) switches off;
(c) The timing sequence begins.

Pin 4 (Reset) This pin is used to reset the device and return the output to a low state—terminating any timing sequence that may be in operation. When not used in a circuit, this pin should be connected to V+ to avoid any possibility of false resetting.

Pin 5 (Control voltage) This pin is connected to the $\frac{2}{3}$ V+ voltage-divider point, which is the reference level for the upper comparator. In most applications of the 555 timer this pin is not required and in these cases should be connected to earth for immunity to noise via a 0.01 microfarad capacitor, as it is a comparator input.

Pin 6 (Threshold) This pin is the external input to the upper comparator (the other being the $\frac{2}{3}$ V+ reference point) and is used to reset the RS FF, which causes the output to go low. When an input greater than $\frac{2}{3}$ V+ is applied, the reset action takes place. Because the A_v of the comparator is high it is very sensitive to the voltage

difference across its two inputs. This allows a slow rate of change (say from a charging resistor-capacitor combination) to increase until just over the $\frac{2}{3}$ V+ voltage level, when the comparator output goes into a logic **1** condition and resets the RS FF.

Pin 7 (Discharge) This pin connects to the open collector of an NPN BJT, the emitter of which connects to negative, internally. The BJT is biased 'on' when the output is low, so that at the end of a timing period it can be used to discharge a timing capacitor. When the output goes high the BJT is biased 'off' and becomes a virtual open circuit as far as pin 7 to earth is concerned.

Unit 25 **SUMMARY**

- A timer is a device which can control on-off timing periods from milliseconds to hours and is also able to provide continuous pulses.
- A very widely used timer is one called a 555.
- The 555 timer has, internally, a voltage reference voltage divider, two comparators, a resettable RS FF, a capacitor discharge circuit and an inverting buffer output.
- A 555 timer can operate from 4.5 volts to 16 volts without any appreciable change in operation.

- The pins of an 8-pin DIL packaged 555 are called:
  1. Negative
  2. Trigger
  3. Output
  4. Reset
  5. Control voltage
  6. Threshold
  7. Discharge
  8. Positive

Timers 2—using the 555 timer

26.1 Operating modes

The 555 timer can operate in two basic modes: the *monostable* (sometimes called *one-shot*) mode and the *astable* (*free running*) one.

Monostable means that a device or circuit has one normal, or stable state (mono is from the Greek, meaning 'single'), and another unstable state. A circuit is normally in a stable state and is then triggered to its unstable state for some period of time, after which it reverts to the stable state. (This is why the term 'one-shot' is used: after being triggered it goes to the unstable state for a certain timed period, once only, then back to the stable state again.)

An astable device or circuit has no stable state (the 'a-' coming from the Greek, meaning 'not') and can be in either of two states. When an astable circuit is switched 'on' it can switch between the two states continuously until switched off or disabled in some way. Because of this continuous oscillation we say the circuit is 'free running'.

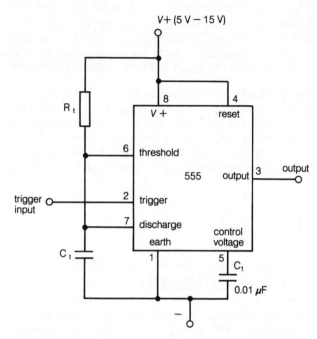

Fig. 26.1 *The 555 timer IC connected as a simple monostable timer*

26.2 The monostable RC timer

The most basic mode of operation of the 555 timer is as a triggered monostable timer. The connections required for this condition of operation can be seen in Figure 26.1. The first impression of the circuit is of its basic simplicity: it consists of the 555 timer IC and the RC timing circuit, a resistor, and capacitor (R_t and C_t). A second capacitor, C_2, connected between pin 5 and earth, is optional. It plays no part in the circuit operation but is included for noise immunity as outlined in Section 25.3, because noise may cause false triggering.

The supply voltage, within the range shown, is connected between pin 8 (positive) and pin 1 (negative or earth). Note also that pin 4 (reset) is also connected to the positive supply since it has no use in this circuit. Pin 4 requires a low input to effect a reset action—so permanently connecting it to a high V+ completely nullifies it.

The output can be connected to any device or circuit which will source or sink up to 200 milliamperes. Additionally, for indication it could be connected to a small indicating lamp or LED. When the output goes high during a timing period, the lamp or LED would light, indicating this condition. (The indicating lamp or LED would be connected between the output and earth.)

Pin 2 (the trigger input) controls the onset of the timing period. Remember that this pin is connected internally to the lower comparator inverting input, and the non-inverting input of the lower comparator is at

$\frac{1}{3}$ V+. If the voltage at the trigger input falls below this reference, the internal RS FF will latch and its output will go low. This will cause the 555 timer output to go high. Once the RS FF latches, the trigger has no effect until the end of the timing period. The trigger input pulse must last no longer than the timing period or the timer will start another timing period. If connected to TTL circuitry, operating at its normal 5 volts, a logic **0** output will operate the trigger, as it is less than 0.4 volt and well below the $\frac{1}{3}$ of 5 volts (1.6 volts) required for triggering.

Pin 6 (threshold) is connected to the upper and more positive side of the capacitor. When the voltage at this point reaches $\frac{2}{3}$ V+ the RS FF latch will release and the output of the 555 timer will go low. This is the basis of monostable operation.

Pin 7 (discharge) is connected to the same point as pin 6. The purpose of pin 7 is to discharge the capacitor at the end of the timing period, ensuring that the timing period will be constant for the same RC combination.

Let us now study the timing operation of the monostable RC timing circuit. Referring to Figure 26.2, which is a timing diagram, it can be seen that initially after supply is connected to the circuit the output is low. Also, at this time both the capacitor and the threshold input (connected to the same point) are connected to earth through the conducting internal BJT.

When the trigger input falls below $\frac{1}{3}$ V+, the internal RS FF latches and the output goes high. At the same time the internal capacitor discharge BJT is biased off and the capacitor C_t commences to charge through R_t.

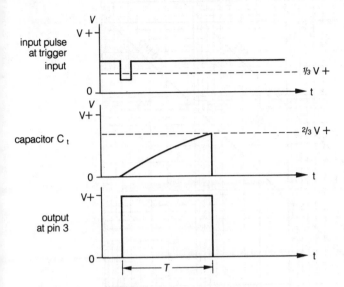

Fig. 26.2 *Timing diagram for a 555 monostable timer*

The capacitor will continue to charge until the voltage across it reaches the threshold voltage, $\frac{2}{3}$ V+, at which point the internal RS FF latch will release, the 555 timer output will go low, and the capacitor discharge circuit will switch on to discharge capacitor C_t. This is the end of the timing period. This period is shown as T on the timing diagram. The circuit is now back to its stable state and awaits another timing pulse to go into its unstable timing stage again.

26.3 Using a monostable timer

Consider once more the circuit in Figure 26.1. The values of both R_t and C_t affect the timing period: the higher the resistance of R_t the longer it will take to charge C_t; the higher the capacitance of C_t the longer it will take to charge. For this circuit the timing period, T, can be calculated from the equation:

$$T = 1.1\,RC \qquad (26.1)$$

where T is timing period in seconds,
R is resistance of timing resistor, and
C is capacitance of timing capacitor.

Example 26.1

If in the circuit in Figure 26.1 the capacitor C_t has a capacitance of 1 microfarad and the resistor R_t a resistance of 4.7 megohms, what will be the timing period?

$$R = 4.7\,M\Omega$$
$$C = 1\,\mu F$$
$$T = ?$$
$$
\begin{aligned}
T &= 1.1\,RC \\
  &= 1.1 \times 4.7 \times 10^6 \times 1 \times 10^{-6} \\
  &= 5.17\ s
\end{aligned}
$$

Answer: The timing period will be 5.17 seconds.

However, to be realistic, the timing period may be 10 per cent either side of this figure because capacitors usually have at least this tolerance. What would be more practical would be to use say a 5 megohm variable (or preset) resistor, and adjust it until the timing period was exactly that required. This does not reflect on the 555 timer, just on the usual wide tolerance of the average capacitor. Of course, lower tolerance capacitors (1 per cent) are available, but are usually more difficult to obtain and more expensive.

Rather than bother to calculate the capacitor and/ or resistor value to determine the timing period, use may be made of the nomograph in Figure 26.3. Note that this is a logarithmic graph and the decade spaces (10 to 100, 100 to 1000 etc.) are equal, but the digits between them are not. If, for example, a 3.3 microfarad capacitor was chosen and a timing period of say 2.5 seconds required, we could find the intersection of the lines representing these values on the graph.

Now this point falls between two diagonal lines representing resistance values of 100 kilohms and 1 megohm. It can be seen to be closer to the 1 megohm line (actually it is a value of about 680 kilohms), so a variable resistor of 1 megohm could be chosen and varied until the timing period was 2.5 seconds. If any two of the values of time, resistance and capacitance are known, the other unknown may be determined from the nomograph.

A practical circuit using the 555 timer IC can be seen in Figure 26.4. This is a motor starter timer circuit and is referred to as a *power-up one-shot* timing circuit. In operation, when power is applied to the primary of the transformer, the half-wave rectifier and shunt regulator supply 12 volts to the timing circuit. The trigger voltage at pin 2 at this time is low, because at the commencement of the charging period the current flowing into the capacitor is a maximum and thus most of the supply voltage is dropped across R. This immediately causes the internal RS FF to latch and the output to go high. This operates the relay connected to the output. Now, as the capacitor, C, charges through resistor, R, its voltage increases until at $\frac{2}{3}$ V+ it reaches the threshold level (at pin 6) and releases the latch on the internal RS FF. The output then goes low and the relay connected to the output is de-energised. By adjustment of R, the timing period can be from practically zero to a maximum of about 10 seconds.

26.4 The astable RC circuit

As we have seen in Section 26.1, an astable circuit is a 'free-running' one which switches continuously between high and low outputs. As for a simple monostable circuit, the basic astable one is not at all complex and only needs the addition of one resistor more than the monostable type. This basic circuit is seen in Figure 26.5. The three timing components are R_{ta}, R_{tb} and C_t. All the other connections to the 555 timer IC are the same as the monostable circuit except that pin 7 (discharge) now

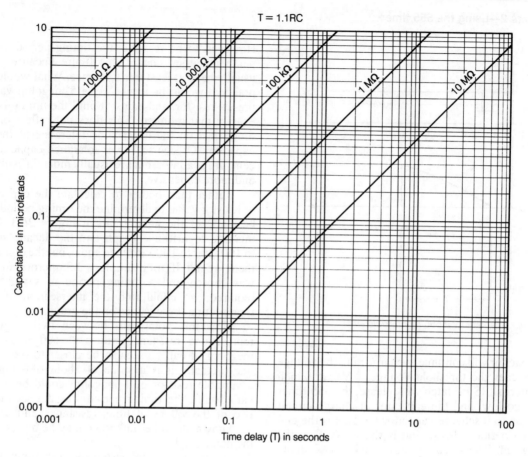

Fig. 26.3 *A nomograph for determining circuit values for timing periods for the circuit of Figure 26.1*

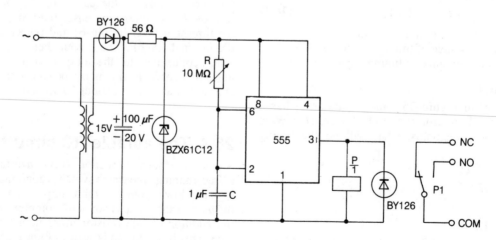

Fig. 26.4 *A motor starter timing circuit based on a 555 timer IC*

connects to the junction between the resistors R_{ta} and R_{tb}.

The circuit goes into operation as soon as the supply is connected, since the trigger voltage is low due to practically all the supply voltage being dropped across R_{ta} and R_{tb}. The internal RS FF latches, and the output goes high. The internal discharge BJT is also biased off. Capacitor C_t commences to charge and when the voltage across it reaches $\frac{2}{3}$ V+ (threshold level), the internal RS FF is unlatched and the output goes low.

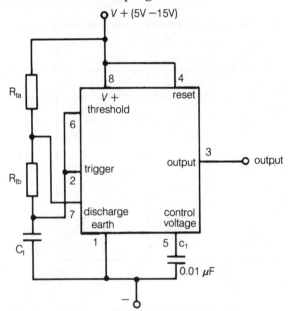

Fig. 26.5 A 555 timer IC connected as a free running astable multivibrator

This then biases on the internal discharge BJT and the capacitor, C_t, discharges through R_{tb}. When the voltage across capacitor C_t reaches $\frac{1}{3}$ V+ (trigger level) the internal RS FF once more is latched and the output goes high. The internal discharge BJT is biased off and capacitor C_t once more commences to charge. This process is continuous until the supply is removed. Quite often, the 555 astable mode circuit is referred to as a *multivibrator* circuit. This is because it effectively 'vibrates' between two limits indefinitely (while supply is still connected, of course).

This process may be seen by referring to Figure 26.6. Here the upper curve represents the voltage across the capacitor C_t in Figure 26.4, and is sometimes referred to as the timing ramp. It can be seen that it increases up to $\frac{2}{3}$ V+ and then decreases to $\frac{1}{3}$ V+ and oscillates between these two limits. At these two limits the output of the 555 timer switches between high and low, as seen in the lower curve.

The ratio of time taken for the high and low periods, t_1 and t_2, is controlled by the ratio of the resistance of R_{ta} and R_{tb}. If R_{ta} is much smaller than R_{tb} the ratio will be approximately 1 : 1.

The time taken for a complete cycle, T, can be determined from the equation:

$$T = 0.693 \, (R_{ta} + 2R_{tb}) \, C_t \qquad (26.2)$$

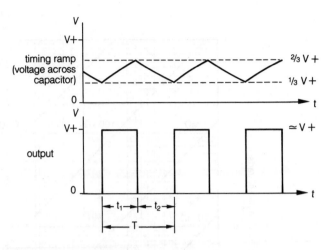

Fig. 26.6 Timing diagram for a 555 astable multivibrator

where R_{ta} is resistance of upper timing resistor,
R_{tb} is resistance of lower timing resistor,
C_t is the capacity of the timing capacitor, and
T is the periodic time.

Now as $f = \dfrac{1}{T}$

where f is frequency, and
T is periodic time,

Equation 26.2 may be re-written as:

$$f = \frac{1.44}{(R_{ta} + 2R_{tb}) \, C_t} \qquad (26.3)$$

Example 26.2

If, in the circuit of Figure 26.5, the values of the timing components were as follows:

R_{ta} has a resistance of 1.2 megohms,
R_{tb} has a resistance of 1 kilohm,
C_t has a capacitance of 0.1 microfarad,

what is the frequency of operation?

$$R_{ta} = 1.2 \text{ M}\Omega$$
$$R_{tb} = 1 \text{ k}\Omega$$
$$C_t = 0.1 \ \mu\text{F}$$
$$f = ?$$

$$\begin{aligned}
f &= \frac{1.44}{(R_{ta} + 2R_{tb}) \, C_t} \\[4pt]
&= \frac{1.44}{(1.2 \times 10^6 + 2 \times 10^3) \times 0.1 \times 10^{-6}} \\[4pt]
&= 11.98 \text{ Hz}
\end{aligned}$$

Answer: The frequency of oscillation is approximately 12 hertz.

Rather than use equation 26.3 to determine the frequency of oscillation of a simple astable 555 circuit, use can be made of the nomograph in Figure 26.7. This

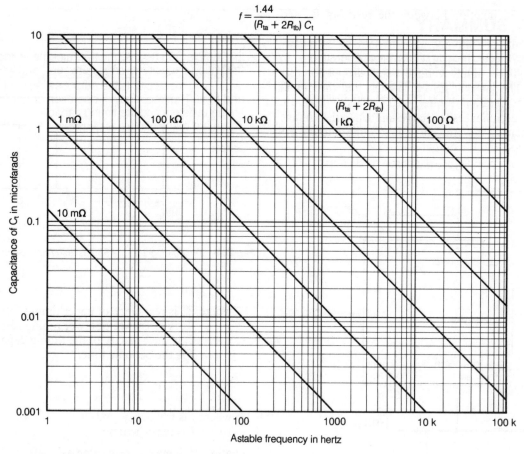

$$f = \frac{1.44}{(R_{ta} + 2R_{tb})\,C_t}$$

Fig. 26.7   *A nomograph for determining the operating frequency of a simple astable multivibrator from timing circuit values for the circuit in Figure 26.5*

is used in a similar manner to the nomograph in Figure 26.3 and also uses a logarithmic scale. The diagonal lines represent the numerator identity $(R_{ta} + 2R_{tb})$ in equation 26.3.

26.5 A practical astable circuit

The 555 timer can be used in many hundreds of different circuits, probably far exceeding the imagination of the people who originated the IC. We will look at only one use of the 555 as an astable oscillator.

Figure 26.8 shows the circuit of an audible continuity tester. The 555 timer is connected as an astable multivibrator, which will give a clearly audible signal when the probe leads are connected across a closed circuit.

This could be used when checking continuity of windings or identifying ends of windings. Another use could be for identifying cables, where the cable is temporarily earthed at one end of a run and the tester connected between earth and the cable at the other. The speaker could be a very small unit salvaged from a defunct transistor radio.

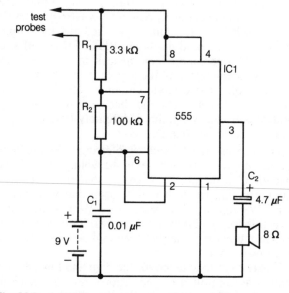

Fig. 26.8   *A 555 timer used in an audible continuity tester*

Unit 26 **SUMMARY**

- There are two basic modes of 555 timer operation: monostable and astable. Monostable is also called 'one-shot' and astable 'free-running', or 'multivibrator'.
- A monostable timer must be triggered into its timing period by a low pulse at the trigger input terminal.
- The timing period of a monostable timer is completed when the voltage at the threshold terminal is $\frac{2}{3}$ of the V+ supply voltage.
- The output of a 555 timer will source or sink up to 200 milliamperes.

- In a simple 555 astable multivibrator the voltage across the timing capacitor swings between $\frac{1}{3}$ and $\frac{2}{3}$ of the V+ voltage. At these points the output swings between low and high. The frequency of operation depends on the timing resistors and the timing capacitor.
- There are many hundreds of uses for a 555 timer in both monostable and astable modes.

Optoelectronics 1— light and electronics

27.1 Light and colour

Electronic devices which use or respond to light in their operation are termed *optoelectronic devices*. Also, in many cases these devices respond to electromagnetic radiation which lies outside the accepted frequencies which we refer to as light.

Light, in its accepted sense, is the *electromagnetic radiation which stimulates the eye*. This is only a very small part of the accepted electromagnetic spectrum, as can be seen in the chart in Figure 27.1.

The frequency (and wavelength) of light falls between two areas of radiation. These are *infrared* (i.e. *less* than red) and *ultraviolet* (i.e. *more* than violet). Neither of these is visible to the human eye, but sunlight contains both ultraviolet and infrared radiation. Also, many optoelectronic devices (and photographic film) are activated by both infrared and ultraviolet rays.

Light is composed of varying wavelengths, from red at the lower frequency end, to violet at the upper frequency. The relationship between frequency and wavelength is given by the equation:

$$f = \frac{v}{\lambda} \qquad (27.1)$$

where f is frequency in hertz,
v is velocity of electromagnetic radiation, (2.9979×10^8 metres per second), and
λ is wavelength in metres (λ is the Greek lower case letter *lambda*).

Example 27.1

What is the wavelength of light that has a frequency of 5×10^{14} hertz?

$f = 5 \times 10^{14}$ Hz

$v = 3 \times 10^8$ ms^{-1} (approximated to a realistic figure)

$\lambda = ?$

$$\lambda = \frac{v}{f}$$
$$= \frac{3 \times 10^8}{5 \times 10^{14}}$$
$$= 0.6 \times 10^{-6} \text{ m}$$

Answer: The wavelength is 0.6×10^{-6} metres. Note: it is usual to quote the wavelength of light in nanometres (10^{-9} metres) so we can call the answer 600 nanometres (abbreviated to nm).

Figure 27.2 shows the light spectrum, together with infrared and ultraviolet radiation, with the wavelength in nanometres and also the frequency in hertz. It can

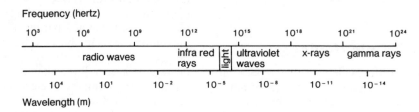

Fig. 27.1 *The electromagnetic spectrum from low frequency radio waves to gamma rays*

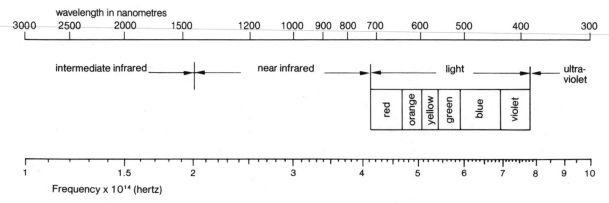

Fig. 27.2 *Infrared, light and ultraviolet spectrum*

be seen that light falls between about 700 and 400 nanometres.

The human eye, while responding to all wavelengths in the visible spectrum, responds best to a wavelength of about 550 nanometres (5.4×10^{14} hertz), which corresponds to a yellow colour. The spectral response of the human eye can be seen in Figure 27.3 where response stimulus is plotted against wavelength.

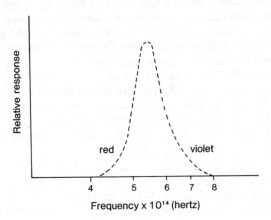

Fig. 27.3 *Spectral response of the human eye*

Sunlight contains all the wavelengths of light and is called *white light*. Different light sources mix the wavelengths in differing proportions, so what we call white light may take on slightly differing appearances—for example light from an incandescent lamp, and sunlight.

27.2 Optoelectronic devices

Optoelectronic devices differ widely in their operation, use and application, and they only share one common feature—all operate as a consequence of the effect of light.

Table 27.1 lists the various optoelectronic devices covered in this book.

27.3 Photodiodes

As discussed in Section 2.9, a depletion layer is formed at a PN junction. When reverse biased, a very small current flows because of electron-hole pairing produced by heat energy. This electron-hole pairing and consequent reverse leakage current will increase rapidly with increase in temperature.

Heat energy is a form of elecromagnetic radiation, and light energy is electromagnetic radiation at a slightly higher frequency (see Figure 27.2). Therefore, light falling on the junction will have the same effect as heat, and produce electron-hole pairing. These holes and electrons, formed by the action of light, are acted on by the reverse bias potential, which sweeps the holes one way and electrons the other, and constitutes a current flow in the reverse bias direction. The intensity of the current flow is proportional to the light intensity at the junction.

Photodiodes are constructed with a small window in one end of the encapsulation. Quite often this is in the form of a miniature lens which directs parallel rays of light onto the PN junction. This effect can be seen in Figure 27.4, which also shows the symbol for a photodiode. Like all semiconductor device symbols this may be inverted or mirror reversed without affecting its meaning.

The reverse bias current which flows in a photodiode as a result of illuminance at its PN junction is quite small and is in the order of about 20 nanoamperes per lux. Figure 27.5 depicts the illuminance characteristics of a typical photodiode. The dark current would be the normal reverse bias current of any diode. As the illuminance level increases, the current also increases, and it can be seen that an increase in voltage only shows a small increase in current at a particular illuminance. A typical maximum reverse bias voltage allowed across a photodiode is about 20 volts, and the diodes usually operate at a very low voltage. This is because there is a limit to the numbers of electron-holes pairs produced at a particular illuminance.

Because the reverse currents of photodiodes are so very small, they must be amplified before they can be used. In Figure 27.5 a photodiode is used with a BJT

Table 27.1 *Optoelectronic devices*

Classification	Device	Effect
Light controlled	Photodiode Phototransistor	Light controls current Light controls current
Light emitting	Light emitting diode (LED)	Light emitted
Light coupled	Optocoupler	Light used as coupling between circuits
Photoconductive	Light dependent resistor (LDR)	Light effects conductivity
Photovoltaic	Solar cells	Light produces emf
Light triggered	Light activated SCR (LASCR)	Light used to trigger into conduction

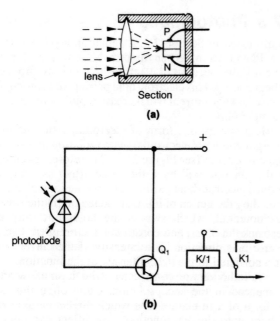

Section

(a)

(b)

Fig. 27.4 *(a) Section through a photodiode, and (b) a photodiode light relay. Note the symbol used for a photodiode and the manner in which the light is focused on the PN junction in the section view*

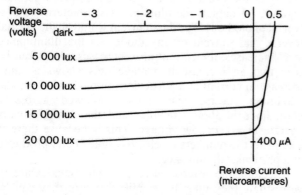

Fig. 27.5 *Illuminance characteristics of a photodiode*

to operate a relay. At a certain light level the base current, produced by light falling on the reverse biased PN junction of the photodiode, will allow sufficient collector current to flow and operate the relay.

A more elaborate and practical example—for a flame failure detection circuit in an oil-fired furnace—is seen in Figure 27.6. If for some reason the flame in an oil-fired furnace is extinguished, the vaporised fuel would still be blown into the furnace. This build up of unburnt fuel could cause an explosion if allowed to continue. To prevent this a photodiode is set so that the light from the flame falls on its lens. This light will produce reverse current in the photodiode and lower the voltage at the base of Q_1. This in turn will reduce the collector current of Q_1 and increase the positive potential at the junction of R_2 and R_3. This will increase the base current to Q_2 and its increased collector current will operate the relay, $\frac{K}{1}$. This will close the contact K_1, which in turn controls the oil supply solenoid valve.

Flame failure will bias on Q_1 which will bias off Q_2 and de-energise the relay K. This will close the oil supply solenoid valve. When starting the furnace, the control is bypassed and the flame can be monitored manually until it is established.

27.4 Phototransistors

The phototransistor is the equivalent to the combination of a photodiode and a BJT, the diode being represented by the base-emitter junction. Phototransistors are enclosed in the same way as smaller sized normal transistors, except that they have a transparent plastic or glass lens in place of the normal opaque plastic or metal cover.

The collector characteristics of a typical phototransistor, as shown in Figure 27.7, are similar to those of a normal BJT (see Figure 6.9)—with illuminance units taking the place of base current. In many cases the base is not connected in the circuit, but if so it is reverse biased so that collector current is completely cut off under dark conditions. The action of the light sweeps

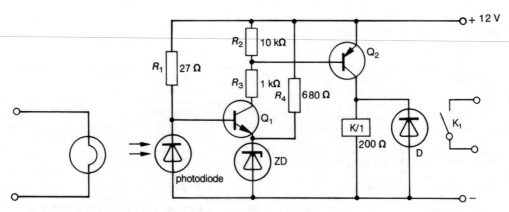

Fig. 27.6 *Photodiode-actuated light detector, for flame failure detection*

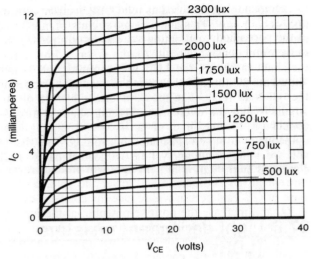

Fig. 27.7 *Collector characteristics of a phototransistor. Each curve is for illuminance level and not the base current as in a normal BJT*

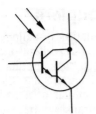

Fig. 27.9 *Symbol of a photo Darlington transistor*

The spectral response of a phototransistor is shown in Figure 27.10. This is identical with the spectral response of the photodiode. It can be seen that the response is mostly in the infrared region, but as the output of a tungsten lamp is also largely in this region, it can be used with very good effect to activate phototransistors and photodiodes.

current carriers across the base, producing a collector current proportional to illuminance.

Phototransistors are used in sound reproduction from movie films and in linear lightmeters. They can also be used for furnace flame failure-control, similar to the photodiode. The circuit for this duty is shown in Figure 27.8. Note the symbol for a phototransistor on the left, and that the base is left open circuit. As this is a purely on-off application, and as the illuminance level is high for active operation, it is not necessary to limit dark current. The action of the circuit is virtually identical for that of the photodiode in Figure 27.6.

To simplify circuits, phototransistors are combined with BJTs to form Darlington pairs. These combinations are termed *photo Darlington transistors*. They are manufactured on a single silicon chip in the fashion of an IC and have very high current gains, which could be up to 10 000. The symbol for a photo Darlington transistor is shown in Figure 27.9.

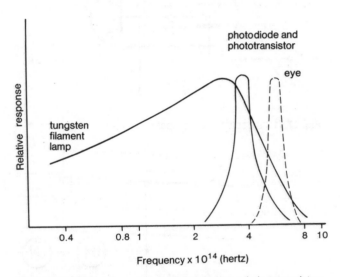

Fig. 27.10 *Spectral response of photodiodes and phototransistors together with the response of the human eye and emitted radiation from a tungsten filament lamp*

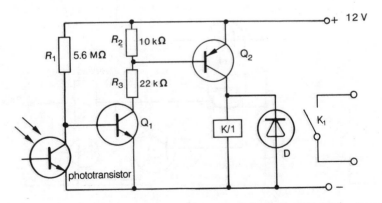

Fig. 27.8 *Flame-failure detection circuit using a phototransistor*

27.5 Optocouplers

Optocouplers (sometimes called optically coupled isolators) provide an excellent means of transferring information between two circuits electrically isolated from each other. They are usually packaged in an 8-pin DIL package, similar to a 555 timer, and consist of a LED and phototransistor optically coupled. The leads to the LED and phototransistor are well isolated and insulated from each other to withstand high voltages. The input-to-output electrical isolation could be up to 7.5 kilovolts. The response time of optocouplers is virtually instantaneous, and they are much more reliable than relays due to immunity from shock and vibration. Their small size is also a decided advantage. A schematic diagram of an optocoupler is shown in Figure 27.11. From this it can be seen that the device is a combination of a LED and phototransistor.

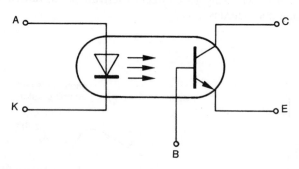

Fig. 27.11 *An optocoupler symbol*

Optocouplers are used as relays, overcurrent sensors and for coupling TTL stages in logic circuits. Figure 27.12 shows a typical application of an optocoupler in an overcurrent sensing circuit. As the voltage drop across the shunt increases, the LED section emits light which allows the phototransistor section to conduct and operate the overload relay. The most important feature of this circuit is that the overcurrent sensing shunt is directly connected to one line of a 415 volt circuit. The optocoupler completely isolates this voltage from the 24 volt operating circuit and yet the overload signal is safely conveyed by the action of light.

27.6 Light dependent resistors (LDRs)

A *light dependent resistor* (also known as an *LDR*) is made from a material (such as cadmium sulphide) which will allow electron-hole pairing when light falls on the surface. The material is like any semiconductor, which at absolute zero is a complete insulator, but in the presence of energy in the form of heat (and in this case light) becomes a conductor. This is because the energy releases electrons and holes to act as current carriers. The LDR is made from cadmium sulphide powder and an inert binding material. This is sintered at a high temperature under carefully controlled conditions. Electrodes, in the form of interleaving combs, are evaporated on to the surface of one side and leads are connected. The completed disc

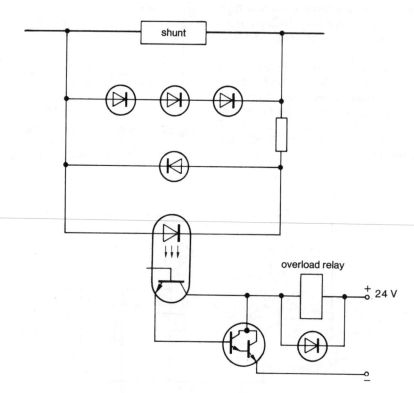

Fig. 27.12 *Photocoupler used as a link between a current overload detector and the overload relay*

is then mounted into an evacuated glass bulb or is encapsulated in transparent plastic to prevent surface contamination. The comb arrangement and the symbol for an LDR can be seen in Figure 27.13 and three typical LDRs are shown in Figure 27.14.

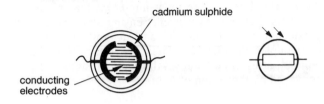

Fig. 27.13 *Sensitive surface and symbol of an LDR*

Alternatively to cadmium sulphide, LDRs can be made from cadmium selenide. The selenide types have a narrower spectral response and are more sensitive to infrared energy rather than light. The spectral response of both types is shown in Figure 27.15 together with the response of the human eye.

27.7 Characteristics of the LDR

The LDR is usually used in on-off applications—that is, when light falls on the surface the result is to either turn on or off some control or process. When an examination of the curve of resistance against illuminance is made (see Figure 27.16) we can see it would be difficult to use the device as a gradual control—but this is possible if only the 'knee' portion of the curve is used. This control is the basis of automatic exposure control in cameras.

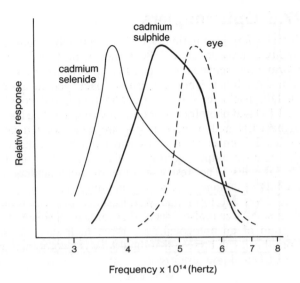

Fig. 27.15 *Spectral response of LDRs compared to the human eye*

As LDRs may be destroyed by overheating, it is very important to keep the power dissipation in the sensitive material below certain maximum values. If the LDR is used to control a circuit in which the power requirements are greater than the LDR can stand, it must operate in a relay circuit or be used as a control for a transistor amplifier. A typical maximum power dissipation would be about 0.5 watt.

The LDR has a time lag between changes in illuminance and changes in resistance—termed its *resistance recovery rate*. The time required to lower its resistance when light is applied after total darkness is

Fig. 27.14 *Three LDRs including a miniature device for camera or lightmeter applications. All can be used in the range 100 to 1000 lux*
PHILIPS ELECTRONIC COMPONENTS AND MATERIALS

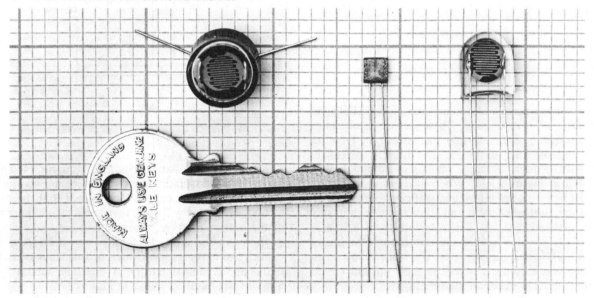

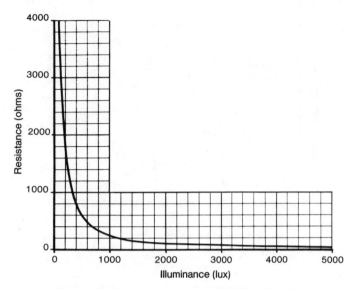

Fig. 27.16   *Resistance illuminance characteristics of an LDR*

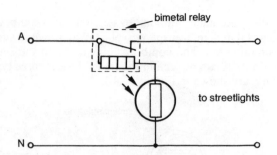

Fig. 27.18   *Streetlighting twilight switch*

very short, usually just a few milliseconds. The rise in resistance with the complete removal of light is much slower and could be up to one second. For this reason it could not be used where very rapid fluctuations of light must operate control equipment.

The resistance recovery rate of a typical LDR is shown in Figure 27.17. This shows the quite quick drop in resistance when light is applied and the rather slower rise in resistance when light is totally removed.

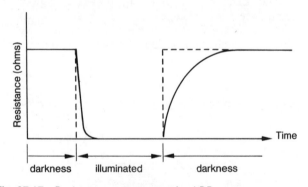

Fig. 27.17   *Resistance recovery rate of an LDR*

One of the more widespread uses of the LDR is in the control of streetlighting. These devices, sometimes called *twilight switches*, switch the streetlight on after sunset and off after daybreak. The LDR is placed in series with the high resistance winding of a bimetal relay, as shown in Figure 27.18. When daylight is falling on the LDR its resistance is low, and sufficient current flows to heat the bimetal element and allow the switch to stay in the open position. As darkness falls, the resistance rises to a high value and the current through the heater element falls to a negligible value. The heater cools down and the switch contacts revert to their normal closed position, which switches on the streetlights.

27.8 Solar cells

Solar cells consist of a thin wafer of N-type silicon, with a substance such as boron, which produces a very thin P-type layer, diffused on to one surface. Metal is plated on the lower side of the silicon and a collector ring on the P-type surface.

When light falls on the surface of the P-type layer, the energy penetrates to the junction and releases hole-electron pairs. The electrons move to the N-type base and the holes to the P-type layer. This produces a positive potential at the collector ring and a negative potential at the base.

The open circuit voltage of a solar cell illuminated by 20 000 lux (sunlight level) and above, is about 0.6 volt and will deliver about 300 milliamperes for each 10 square centimetres of surface area. Batteries of solar cells are used to power satellites and space vehicles and some rural telephone services. A section through a silicon solar cell is shown in Figure 27.19 and an array of silicon solar cells in Figure 27.20.

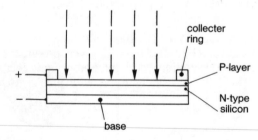

Fig. 27.19   *Section through a silicon photovoltaic cell*

When used for battery charging, as in rural telephone exchanges and isolated microwave repeater stations, the design of the solar cell battery is such that the energy on an average day will more than supply the requirements of the battery load. As charging can only take place in daylight (and on sunny days) the battery must have ample capacity to carry over for three or four days of sunlight of no appreciable intensity. A charging circuit for an array of solar cells is shown in Figure 27.21.

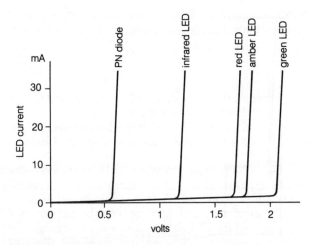

Fig. 27.22 *Characteristic curves of diodes and LEDs*

Fig. 27.20 *Use of silicon solar cells to charge the battery of a yacht by the action of sunlight* ELCOMA, A DIVISION OF PHILIPS ELECTRICAL INDUSTRIES PTY LTD

27.9 LEDs (light emitting diodes)

As mentioned in Section 2.11, LEDs are diodes which emit light at the junction when current is flowing through them. In a silicon (or germanium) diode the losses at the junction are represented by heat, but when certain other materials (such as the compound gallium arsenide) are used to form PN junctions, the losses are represented as light.

The voltage drop across LEDs is much higher than the voltage drop across silicon diodes (see Figure 27.22) and LEDs thus have a lower efficiency than silicon diodes. However, the very fact that they operate at a lower efficiency allows us to use the losses as visible light.

LEDs use doped gallium arsenide to vary the wavelength of the emitted light. LEDs are produced which emit infrared radiation, red light, amber light and green light. The LED is usually fitted with a lens to direct the light in a narrow or wide beam depending on its application.

When LEDs are used as indicators, a dropping resistor must be connected in series to keep the LED current at a safe level. The typical indicator LED circuit is shown in Figure 27.23. Note that the LED symbol is different from all the other optoelectronic devices covered so far in this unit in that the arrows, signifying the action of light, are directed *from* the symbol and not to it. The LED *emits* light, rather than responding to the action of light.

The current requirement of a LED to produce its designed light output may vary from 17 milliamperes

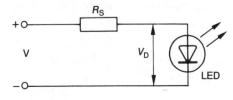

Fig. 27.23 *A LED used as an indicator*

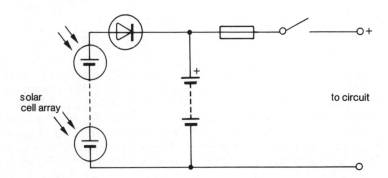

Fig. 27.21 *Solar cell battery charging circuit*

to 50 milliamperes. When connected to a dc supply the value of the series resistor can be calculated as follows:

$$R_S = \frac{V - V_D}{I} \qquad (27.2)$$

where R_S is the resistance of the series resistor,
V is the voltage of the supply,
V_D is the inbuilt voltage drop across the LED,
I is the desired current through the LED.

The voltage drop across the LED varies, depending on the type of LED, but an approximation is usually made and a figure of 1.7 volts used. In addition, it is usually simpler to designate an average current of 20 milliamperes for all LEDs (unless a more accurate light output is required). This greatly simplifies the series resistance calculation.

Example 27.2

Determine the resistance of a series resistor for a LED indicator to be used on a 24 volt dc supply.

$$V = 24 \text{ V}$$
$$V_D = 1.7 \text{ V}$$
$$I = 20 \text{ mA}$$
$$R_S = ?$$

$$R_S = \frac{V - V_D}{I}$$
$$= \frac{24 - 1.7}{20 \times 10^{-3}}$$
$$= 1115 \ \Omega$$

Answer: Use a 1.2 kilohm resistor (nearest preferred E12 value).

Unit 27 **SUMMARY**

- Electronic devices which use, or respond to, light are called optoelectronic devices.
- Light is electromagnetic radiation which stimulates the eye.
- Light is measured in both frequency and wavelength. Wavelength is always in nanometres.
- Photodiodes use the reverse bias current produced by light falling on the PN junction.
- Phototransistors use light falling on the base-emitter junction to control collector current.
- The base of a phototransistor is left either disconnected or reverse biased so that dark current is zero.

- Optocouplers use light as a means of conveying signals between two electrically incompatible circuits.
- The resistance of an LDR is dramatically reduced when light falls on its surface. LDRs are used in twilight switches.
- Solar cells produce electricity from sunlight.
- LEDs emit light from their PN junction when forward biased.
- They must be used in conjunction with a dropping resistor to keep the current through them to about 20 milliamperes.

Optoelectronics 2— displays and control

28.1 The seven-segment display

We have seen in Section 24.5 that indicator LEDs can display binary numbers. This is achieved by having an 'on' LED represent binary **1** and an 'off' LED represent binary **0**. These displays, however, are much more difficult to immediately interpret than the more familiar decimal numbers.

Whereas we can use a single LED to represent either of the two binary digits, it would take ten separate displays to represent the digits of the decimal system. At one time this was done by using special neon tubes with ten anodes, each representing a decimal number. Later, neon tubes with ten superimposed wire anodes in the shape of the decimal digits could display the required numerals. All these have now been completely eclipsed by the *seven-segment display*. Seven-segment displays can be made from various lighting methods but in this section we will examine only the LED display.

By stylising the ten digits in the decimal system it was found that they could be reduced to seven fundamental straight lines, without creating any doubt as to which number the combination of lines represent. This effect was enhanced by placing all vertical lines at an inclined angle of 20^0 to the vertical. Each of the seven lines is termed a *segment* and various combinations of the segments can form an easily recognised representation of the decimal numerals.

When representing a numeral the particular segments required for that numeral are lit, the others being unlit. Each segment, therefore, is a LED which has its own anode and cathode connection. A representation of a seven-segment display showing three different numerals is shown in Figure 28.1.

Each of the seven segments is represented by a letter starting with the letter 'a' at the top and going clockwise to the letter 'f' and then assigning the centre segment the letter 'g'. This is shown in Figure 28.2. In addition

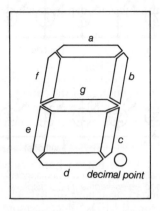

Fig. 28.2 *The segments of a seven-segment display with their designated letters*

there is a decimal point which is a separate LED. Figure 28.3 shows the typical pin-outs of a seven-segment LED display and the segments which connect to each pin.

Seven-segment LED displays can be connected in one of two ways—*common anode* and *common cathode*. The display in Figure 28.3 has two pins marked 'anode'. This is, of course, a common anode type display and the positive connection may be made to either or both of these points. Each of the other active pins must be connected to negative to light its particular segment. In Figure 28.4 the internal schematic diagram of the individual LEDs is shown in both common anode (a) and common cathode (b) connection.

Seven-segment LED displays are available in several styles and digit sizes. They normally range in size from 2.6 millimetres to 15.2 millimetres in the vertical direction. Mostly they use red LEDs but amber ones are also used. The amber colour is most sensitive to the human

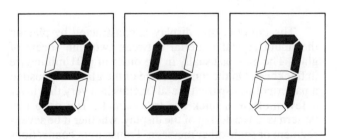

Fig. 28.1 *Any decimal digit can be represented in a seven-segment display*

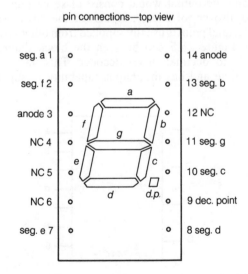

Fig. 28.3 *The pin-outs of an LT302 seven-segment LED display unit*

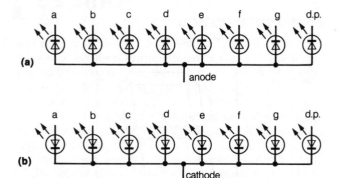

Fig. 28.4 *Internal circuits of (a) common anode, and (b) common cathode seven-segment LED displays*

eye (see Figures 27.2 and 27.3) so amber LEDs have the highest luminous efficacy. One amber seven-segment 12.7 millimetre LED display has a luminous intensity of 3,300 microcandelas for an input of 400 milliwatts. A similar red LED display has an output of only 600 microcandelas for the same input wattage.

In the sizes 7.6 millimetres and 12.7 millimetres, the red LED seven-segment displays require an average current of 25 milliamperes per segment, with a voltage drop of about 1.7 volts.

28.2   The BCD decoder

In order to use a seven-segment LED display to indicate a decimal numeral we must first employ a *binary coded decimal decoder*. This device has internal circuitry which will activate the correct display segments to indicate a decimal digit when the input to the device is a binary numeral. As a single seven-segment display can only indicate the decimal digits 0 to 9, the binary input need only be from binary **0** to binary **1001**. This would indicate that the input to the BCD decoder would consist of four pins and the output would consist of seven pins. There would also be, of course, inputs for the power supply. The decimal point is usually supplied from other circuitry.

In Figure 28.5 can be seen the block diagram of a BCD to decimal driver/decoder. This unit is TTL compatible and has the output capability to supply the

correct current to a seven-segment LED display. The inputs are coded A, B, C, and D and represent the outputs of a counter (see Figure 24.11) which will input its particular logic state, representing a binary digit. The outputs coded a, b, c, d, e, f and g connect to the corresponding inputs on the seven-segment display.

28.3   Liquid crystal displays

Liquid crystals are substances which behave differently from other substances because of the arrangement of their molecules. When an ordinary substance melts, its molecules lose their molecular cohesion and move in all directions. For example, when ice melts it forms water and has completely different properties than when it was ice. The change of state between solid and liquid is very sudden and we say there is an extremely small temperature differential between these two states.

Liquid crystals, on the other hand, do not pass directly from solid to liquid, but form a state intermediate to these two. This state can exist over a wide temperature range and it is notable that the molecules forming the substance can slide past one another only in certain directions. This is because the molecules have a definite polarity.

In its normal state the molecules of a liquid crystal are aligned in the same direction and are transparent. Under the influence of a magnetic field, the molecules become misaligned and change from transparent to reflective, as can be seen in Figure 28.6. This is the basis of the *liquid crystal display*, or *LCD*. The liquid crystal material is said to be *nematic*, that is, it exhibits the properties of a solid (its molecules aligned and in a definite pattern), even though it is in a liquid state. Liquid crystal substances are opaque when under the influence of an electrical field.

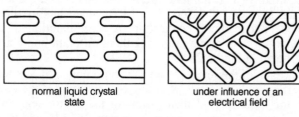

normal liquid crystal   under influence of an
state   electrical field

Fig. 28.6 *Representation of arrangement of molecules in a liquid crystal: in normal state, and when acted on by an electrical field*

The liquid crystal display is constructed by placing the liquid crystal material between two thin sheets of glass. The glass is usually in the order of 0.01 millimetre in thickness. On the inner surface of the glass is deposited a pattern of transparent metal (metals in a very thin layer a few molecules thick are transparent). The deposited pattern is in the design of the display, whether it be seven-segment or some other message. Transparent connections are made to the segment (or other) patterns, and taken to external terminals.

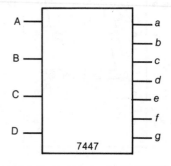

Fig. 28.5 *Input and output terminals of a 7447 BCD decoder/ driver*

The thin glass sheets containing the electrodes and liquid crystal are called the LCD window and on each side of this is placed another sheet of glass. The whole combination is referred to as an *LCD cell*. If both the sheets of outer glass are transparent it is called a *transmissive-type cell*—in this type, light may pass through the cell in any place except where the electrodes have an electric field between them. This is where the liquid crystal has become opaque, and the cell must be backlighted to be effective.

The other type of LCD cell is the *reflective-type cell* and has a reflective layer on the rear glass—light entering the front of the cell is reflected back. However, it cannot pass back through the areas between any electrodes which have been made opaque by an electrical field. These areas appear black when viewed under light. This is the most common type of LCD cell. A representation of the construction of an LCD cell is shown in Figure 28.7.

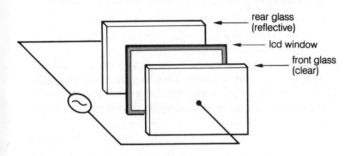

Fig. 28.7 *The simplified assembly of an LCD cell*

It may be noted in this figure that an ac source is connected between each side. There is a very good reason for this: if dc was applied to the electrodes of the LCD cell it might, by electrolysis, cause plating onto and from the electrodes, and quickly render them useless. In very small battery-operated equipment such as hand-held calculators or watches, the dc from the battery is changed to ac by oscillators.

The current-demand of LCD cells is very small and for a seven-segment display is in the order of 6 microamperes. This is why liquid crystal displays are ideal for watches and small calculators. Unlike other optoelectronic devices discussed in this unit and Unit 27, LCD cells are neither light-emitting nor light-operated. They simply become opaque to light, with an extremely low power demand. In most apparatuses the energy requirement for the LCD display is only a very tiny fraction of that for the operation of the rest of the equipment.

The operating frequency of the potential applied to the LCD cell is in the order of 50 hertz and is supplied by a built-in oscillator. To produce the opaqueness in the liquid crystal, only a very small voltage is required, but the LCD cell may be operated up to a voltage of about 30 volts peak-to-peak.

28.4 Fibre optics

The transmission of signals by means of light is as old as history, and ranges from bonfires on a hill to morse code using flashing lights. Later came modulated beams of light which could carry the information of speech and music, and later still laser beams which could do the same thing far more effectively. Two problems with signalling by light are that it will only travel in a straight line, and it is rapidly attentuated by the medium through which it travels (air, water, etc.). Even the light from the sun is attenuated by the earth's atmosphere—more so during say a dust storm.

It is a principle of physics that light is *refracted*, or bent, when passing from a less dense to a more dense medium or vice versa. This is quite apparent when a long rod is placed in the water—the rod appears to bend forward at the point it enters the water. This is because the velocity of light in water is less than in air. The amount of bending that apparently takes place depends on the difference in the properties of the air and water. The amount a ray of light bends, when passing from one medium to another, is indicated by the *refractive index* between those two mediums. The greater the refractive index, the greater is the angle of refraction when light passes from one medium to another. If the angle at which the light passes from one medium to another is equal to (or less than) the angle of refraction, the light instead of passing from one medium to another is reflected from the boundary between the mediums. This effect is termed *total internal reflection*. It can be seen in the sparkle of a multifaceted diamond, which has a very high refractive index, and is also seen in prism mirrors in binoculars and some optical instruments.

The effect of total internal reflection is made use of in optical fibres. Figure 28.8 shows a section of optical fibre cable. This has a core of extremely pure glass with a much higher refractive index than the cladding, which is also made from very pure glass. There is total internal reflection at the boundary between the cable core and cladding, which in effect acts as insulation for the light travelling along the core. The light loss in the core is extremely low and even if the cable is bent the internal reflection allows all the light entering the core to be reflected through it.

When dealing with signals, electricity is just one method of conveniently passing signals from one point to another. Another is to use electromagnetic waves at radio frequencies. Now that we can contain light in an

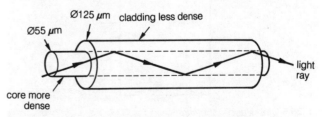

Fig. 28.8 *Total internal reflection in a fibre-optic cable*

optical fibre cable we can make use of light to convey signals. Electricity has the disadvantage of producing a magnetic field which can affect external apparatuses. Moreover, other magnetic fields (by electromagnetic induction) can induce voltages which can degrade signals or even eliminate them. Light, on the other hand, is not affected by any magnetic field, nor does it produce one of its own. This makes the optic fibre an ideal method of transferring a signal immune from electrical or magnetic interference.

The fibre-optic cable has a plastic sheath extruded over the double glass fibre and is terminated by fitting a sleeve over the cable and squaring and polishing the end of the fibre. This can be seen in Figure 28.9 where light which entered the remote end of the cable is seen shining from the polished near end.

Because the frequency of light is much higher than radio frequencies, a very large amount of information can be transmitted in an optic fibre cable. The rate of transmission of digital information is usually expressed in *bauds*, one baud being one code bit per second. Light is capable of conveying information at the rate of 40 megabauds.

Optic fibre cable is finding increasing use in the transmission of signals—the biggest application being in telephone trunk circuits, including undersea cables. However, much use is also being made in industry, where signals for industrial process control are being transmitted by light rather than electricity. Special techniques are required to terminate and join optic fibre cables, making their installation and repair a little more difficult than metallic cables. In many installations, however, the advantages far outweigh the extra effort and cost required.

The fibre-optic cable is interfaced with conventional electrical circuits by using a LED to change electrical signals to light signals and, at the other end, a photo diode to change light signals back to electrical signals. A special terminating module is soldered to a circuit-

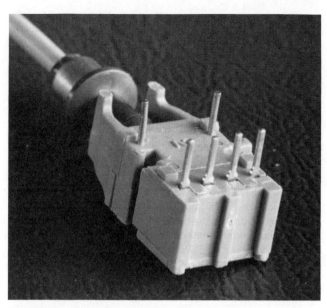

Fig. 28.10 *Fibre-optic cable fitted with an interface module*

board, and the fibre-optic cable with a terminated end is slipped into the module. This can be seen in Figure 28.10. The transmitting and receiving modules are similar and are usually coloured differently, for identification.

28.5 Light activated SCRs

The *light activated SCR*, or LASCR, is similar in construction to the SCR except that a small window—sometimes in the form of a lens—permits light to shine on it, illuminating the silicon pellet to which the gate lead is connected. (See Figure 28.11.) The action of light has the same effect as current flow in the gate. At a certain illuminance level the SCR is triggered into conduction. This can be a purely on-off type of control (when dc is used), or a variable control with ac.

The relatively high-power handling capabilities of the LASCR (typical value 350 watts) allow it to be used to control small power circuits directly. To operate in higher power circuits, the LASCR is sometimes used to trigger a larger SCR.

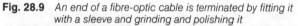

Fig. 28.9 *An end of a fibre-optic cable is terminated by fitting it with a sleeve and grinding and polishing it*

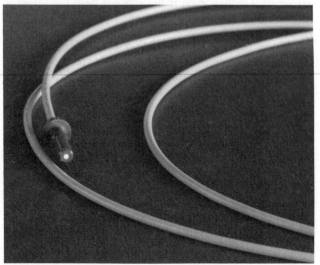

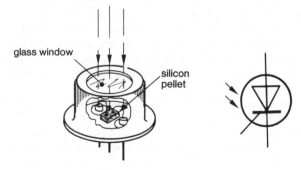

Fig. 28.11 *Light activated silicon controlled rectifier*

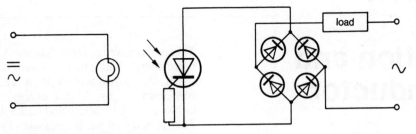

Fig. 28.12 *Relay control using light and an LASCR*

An LASCR may be used as a relay, if the control circuit switches on and off a lamp placed near the window of the LASCR. Any voltage, ac or dc, may be used in conjunction with the lamp, and the circuit is not sensitive to any transient voltage that may be induced in it. Figure 28.12 illustrates such a circuit—note the use of a bridge rectifier circuit to supply a pulsating dc (which can commutate the LASCR when the light goes off) while ac flows through the load. It would be possible to use a fibre-optic cable between the LASCR and light signal source if this was desired.

Unit 28 **SUMMARY**

- Any decimal digit can be represented by a seven-segment display.
- The segments of a seven-segment display are designated a to f, clockwise from the top, and the central segment is designated g.
- Seven-segment LED displays can be common anode or common cathode.
- Each segment of a typical seven-segment LED display draws about 25 milliamperes with a voltage drop of 1.7 volts.
- A BCD decoder supplies four binary digits to represent any decimal digit.
- The output of a BCD decoder/driver will activate the correct segments of a seven-segment display to produce the relevant decimal digit.

- A liquid crystal is in a state intermediate between liquid and solid.
- A liquid crystal becomes opaque under the influence of an electrical field.
- Transparent electrodes, in the form of a seven-segment display, allow an LCD to display any decimal digit.
- An LCD display cell must use ac to activate the electrodes.
- Fibre-optic cables transmit light signals using the effect of total internal reflection.
- Fibre-optic cables are not prone to electrical or magnetic interference.
- LASCRs use light rather than gate current to be triggered into conduction.

Appendix A

Conduction and semiconductor theory

A.1 Electron theory

Matter, which can be defined as anything which occupies space, exists in three forms: solid, liquid, gas.

At the temperatures and pressures normally associated with the human environment, matter usually occupies only one of these forms. A notable exception is water, which may exist as a solid (ice), a liquid, and a gas (water vapour). Matter may be a mixture of different substances (e.g. soil), or it may be a pure substance (e.g. sugar). If it is a pure substance it can be further divided into either *elements* or *compounds*.

An *element* may be defined as 'a substance which cannot be changed into simpler substances by chemical means and which is composed exclusively of similar *atoms*'.

A *compound* is composed of a fixed definite proportion of two or more elements chemically combined. The elements cannot be separated by physical means and a compound can only be broken down by chemical means. A compound may bear no resemblance to its separate constituent elements. For example, the constituent elements of common salt (sodium chloride) are sodium, a dangerous active metal, and chlorine, a poisonous gas.

If a compound or element is subdivided many times, the piece becomes smaller and smaller until it reaches a limit from which it cannot be further divided. This smallest particle is called a *molecule.*

A *molecule* is the smallest particle of a substance that can exist alone and retain the chemical composition of the substance. Molecules are extremely small—too small to be seen by the most powerful optical microscope; a few comparatively gigantic modern plastic or protein molecules may be observed under electron microscopes.

When energy in the form of heat is imparted to molecules they commence to agitate in continuous motion, but when no heat is present, at absolute zero (–273⁰C), all motion ceases. In the solid state, there is little motion because of restriction by the forces of attraction which form the molecules into definite geometrical shapes called *crystals.* As heat is applied, the movement increases and the forces which attract the molecules become less than the energy of motion, thus the molecules are able to move apart more easily. There is still some attraction but not enough to retain the crystalline shape of the solid, so that the substance can take the shape of a containing vessel or spread out under the influence of gravity. This is the liquid form. Further heat energy allows the molecules to move apart and occupy a greater space with no restriction on shape.

This is the gaseous form. Some substances require much more energy than others to change their form.

If the substance is in a confining vessel when converted to gas, further energy causes the molecules to move more violently and strike the walls of the vessel. This is referred to as pressure. Pressure also modifies the temperature at which the change of form from liquid to gas takes place.

All molecules are composed of *atoms*. Some compounds have many atoms in their molecules while some elements, for example helium, have only one.

Atoms of the natural basic elements range from the least dense, hydrogen, to the most dense, uranium. In addition there are the radioactive transuranic artificially produced elements which exceed uranium in density. Hydrogen, in addition to being the least dense atom, is the simplest atom. Uranium is the most complex.

Atoms consist of two major parts—the *nucleus* and the orbiting *electrons*. Electrons are the least dense particles known, and carry a negative charge. The nucleus consists of fundamental particles called *protons* and *neutrons*. The only exception to this is the nucleus of the hydrogen atom which has one proton only (Fig. A.1). Physicists are investigating other subatomic particles but these do not affect the behaviour of atoms in either the chemical or electrical sense.

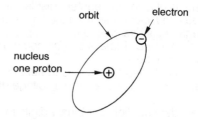

Fig. A.1 *A hydrogen atom*

A *neutron* is a particle with no electric charge but with a mass approximately the same as the proton. An atom of an element may have more or fewer neutrons without affecting its chemical property, only its *atomic weight*. These different atomic weight atoms of the same element are called *isotopes*. Many isotopes are radioactive, that is they emit neutrons from the nucleus; in this way radioactive elements degenerate—for example radium to lead.

In any atom the number of electrons in the orbits equals the number of protons in the nucleus, thus a normal atom has a neutral charge—that is positive charges cancel negative charges. The *atomic number* of an atom is given by the number of either its orbiting electrons or protons in the nucleus, which of course are equal in number. The atomic number of hydrogen therefore is 1.

In order of density the next atom after hydrogen is helium (Fig. A.2). This has two electrons and two protons and, in addition, two neutrons, so the atomic number of helium must be 2. The neutrons appear to be necessary

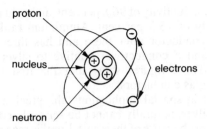

proton

nucleus

electrons

neutron

Fig. A.2 *A helium atom*

to bind the nucleus together, the two positive protons would normally repel each other.

The *atomic weight* of an atom is the total number of protons and neutrons in the nucleus. The atomic weight of hydrogen is 1 and that of helium, 4.

A good example of atomic number and weight can be given by examination of the atom of carbon (Fig. A.3), which has six electrons in two *shells,* two in an inner shell and four in an outer shell. The nucleus consists of six protons (to balance the electrons) and six neutrons, so this carbon atom has an atomic weight of 12 and an atomic number of 6. Other isotopes of carbon exist and these range in atomic weight from 10 to 15. Except for three, however, these isotopes have a life of only a few seconds or minutes, as they are highly radioactive. The isotope carbon 12 accounts for about 98 per cent of all carbon, and that of carbon 13 for just under 2 per cent. The radioactive isotope carbon 14 has a half life of about 5600 years: in other words, from a given sample half will have disintegrated in that time. Scientists usually give carbon an average atomic weight of 12.011

to include the isotopes. The average atomic weight of all the other elements is determined in the same manner.

The shells formed by the orbits of the electrons follow a definite pattern. The innermost shell, called the K shell, has a maximum of two electrons. The next shell, called the L shell, has a maximum of eight electrons and the next outer shell, the M shell, has a maximum of eighteen electrons. In the first eighteen elements (hydrogen to argon) the new electron is always added to the outermost shell until the shell is filled. In the other elements new electrons are added so that there may be two or more unfilled shells. There are never more than eight electrons in the outermost shell. These outermost electrons are called *valence* electrons and are the ones that are of interest in electronics.

Substances with eight valence electrons are extremely stable and show no inclination to enter into chemical combination. (They include the *inert* gases neon, argon, krypton and xenon.) The atoms of most substances exhibit a marked tendency to reach this stable state and can achieve it by sharing electrons in their outer shell with similarly placed electrons of other atoms. This sharing of electrons between two atoms is known as a *covalent bond.*

Between two of the previously discussed atoms, hydrogen and carbon, the covalent bonding of one carbon atom and four hydrogen atoms forms the gas methane (Fig. A.4), also known as marsh gas or fire damp. The sharing of valence electrons between the five atoms produces an outer orbit of eight electrons.

A.2 Conduction and conductors

Elements may be classified into two groups: metals and non-metals. Among other properties, metals are *conductors* of electricity and non-metals (with the exception of carbon) are *insulators* or *non-conductors.* There are, however, some elements which cannot be strictly classified into either group. These are the elements silicon and germanium (which, like carbon, have four outer valence electrons). They are classed as *semiconductors.*

Just as some metals are better conductors of electricity than others, so some non-metals are better

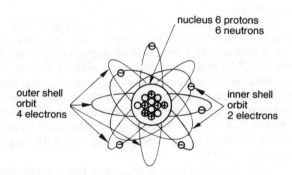

nucleus 6 protons
6 neutrons

outer shell
orbit
4 electrons

inner shell
orbit
2 electrons

Fig. A.3 *A carbon atom*

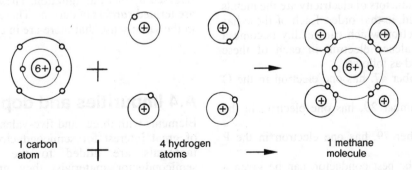

1 carbon
atom

4 hydrogen
atoms

1 methane
molecule

Fig. A.4 *The covalent bonding of hydrogen and carbon forms the gas methane*

insulators than others. The ability of metals to be good conductors and the ability of non-metals to be good insulators depends on the number of valence electrons each has, the distance these electrons are from the nucleus, and the number of electrons in the preceding shell.

Electrical current in a solid is exhibited as a movement of electrons along or through the solid. In insulators the valence electrons have great difficulty in moving from the outer orbit. Even the best insulators have some electron movement, but this is usually so minute that it may be completely ignored.

Metals have very few valence electrons: the better conductors have only one. The single electron in the outer shell is not under such a strong nuclear force as the larger number of outer shell electrons in another conductor. It is relatively simple for this electron to pass from atom to atom within the conductor, so it is said to be in the conduction band or to be a *free electron*.

This motion of the free electrons is normally a random movement, but when a potential difference is applied to the conductor there is a definite movement of the electrons from the negative to positive terminal. The greater the potential difference, the greater the rate of flow of the electrons. It must be understood that the negative terminal is supplying electrons which are being transferred through the conductor and being withdrawn at the positive terminal. At any given time the number of electrons in the conductor is constant. When the supply is of constant polarity, that is dc, the movement of electrons is continuous and in the one direction. When the supply is alternating, the movement of electrons is a back and forth motion.

The flow of electrons constitutes a current and if a current of 1 ampere flows for one second, a charge of one coulomb is carried. This is a flow of 6.3×10^{18} electrons per second. The larger the area of the conductor, the easier it is for these electrons to pass and the lower is their speed. If the cross-section of the conductor is small the speed of the electrons is greatly increased. The speeding electrons affect the thermal state of the atoms as they pass and the agitation of the atoms increases. This makes it more difficult for the electrons to flow through the conductor and it is said that the resistance of the conductor is increased. Whether the heat is supplied by the flow of electrons themselves or by external means, the effect is to increase the thermal agitation of the atoms of the molecules.

The best three conductors of electricity are the metals silver, copper and gold, in that order. Each of these has only one valence electron which can readily become a free electron. The valence electron in each of these conductors is arranged as follows:

silver, atomic number 47, has one electron in the O shell;

copper, atomic number 29, has one electron in the N shell;

gold, atomic number 79, has one electron in the P shell.

Silver, which is the best conductor, can be given a relative conductivity of 100 per cent, copper, the next, a relative conductivity of 96.8 per cent, and gold a relative conductivity of 73.5 per cent. Aluminium ranks as the next best conductor after these and has three valence electrons and a conductivity relative to silver of 59.9 per cent. These figures show that the first three are outstanding as conductors.

Except in special applications, the precious metals, gold and silver, naturally cannot be seriously considered as conductors because of their cost. For this reason copper is at present the premier electrical conductor, although aluminium is finding progressively more use: it has a lower cost than copper and is the better conductor on a weight-to-conductivity basis. Aluminium is used almost exclusively for high voltage transmission lines and is being increasingly used for high current bus bars and as transformer conductors.

As stated previously, all metals are conductors of electricity, but the ones mentioned above are the best.

A.3 Semiconductors

Certain elements which have four valence electrons are known as semiconductors, because while they are not good conductors they are also very poor insulators. The two of most importance in electronics are the elements germanium and silicon. Germanium, in a pure state, has a resistivity of 0.6 ohm-metres and pure silicon a resistivity of 6000 ohm-metres. These values may be compared with the resistivities of copper and mica, which are 1.72×10^{-8} ohm-metres and 10^7 ohm-metres respectively.

If the temperature of these semiconductors is increased, the resistivity is reduced in the same manner as in insulators. At a temperature approaching absolute zero ($-273°C$) their resistivity is as high as that of the insulators. When in the normal state, these materials form into definite crystalline patterns by the *mutual covalent bonding* of the outer four valence electrons. As illustrated in Figure A.4, there is a bonding between each valence electron of an adjacent atom forming the stable eight-electron outer shell.

Figure A.5 represents a section of pure germanium. When a potential is applied to this material, a very small current will flow as a result of electrons and holes being released by thermal agitation. These electrons and holes are termed *carriers* of current. There are no free electrons in the same sense that there are in copper, for example.

A.4 Impurities and doping

Elements with three- and five-valence electrons are also of great interest for semiconductors because, if these materials are added to the pure four-valence semiconductor materials, they greatly modify their electrical characteristics.

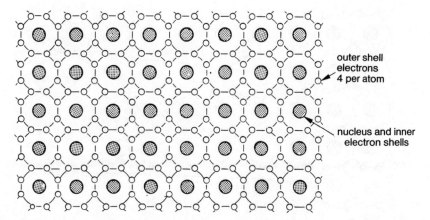

outer shell
electrons
4 per atom

nucleus and inner
electron shells

Fig. A.5   *A crystal matrix of pure four-valence semiconductor. Note how the valence electrons of each atom combine to form an eight-electron outer shell*

Elements with three-valence electrons include boron, aluminium and indium, and those with five-valence electrons include phosphorus, arsenic and antimony.

In semiconductor manufacture the three- and five-valence materials are added to the pure four-valence material in the proportion of about one part in 12 000 000. When this has been carried out the pure germanium or silicon is said to be *doped*.

A.5 N-type semiconductors

When a four-valence semiconductor material is doped with a five-valence impurity, the impurity atoms enter into the crystalline structure forming covalent bonds with four of their valence electrons. The unbonded electron is free to wander through the crystal. This is somewhat similar to the situation existing in metallic conductors.

As electrons are negatively charged and migrate towards the positive terminal if a potential is applied to the crystal, the crystal is said to be composed of *N-type* material. The N stands for *negative current carriers* as it is electrons which carry the current through the crystal.

Figure A.6 illustrates the N-type semiconductor material. Here, three impurity five-valence atoms contribute three free electrons to the structure. It must be remembered, however, that the actual degree of impurities is far less than shown, as stated in Section A.6. The impurity atoms shown in the figure have *donated* electrons to the crystal; they are termed *donor* atoms and the actual impurity element is termed a *donor material.*

A.6 P-type semiconductors

When a four-valence semiconductor is doped with three-valence impurity, the impurity atoms again enter into the crystalline structure, but this time the fourth bond between the four-valence electrons of the semiconductor and the impurity atom cannot be completed. This is said to leave a *hole* in the crystal structure.

This hole in the covalent bond, illustrated in Figure A.7, can readily accept electrons from adjacent atoms. When this happens the hole moves to the place occupied by the electron. When a potential is applied to the crystal, the holes migrate towards the negative terminal to accept

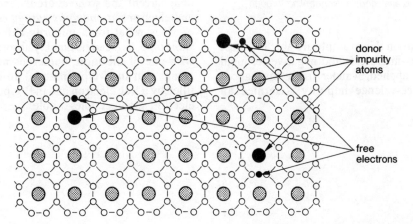

donor
impurity
atoms

free
electrons

Fig. A.6   *N-type semiconductor. The electrons donated by the five-valence impurity act as majority current carriers*

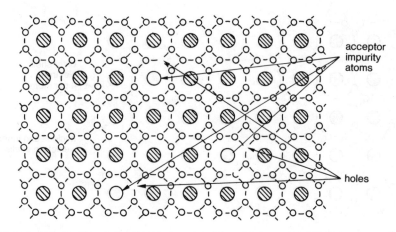

Fig. A.7   *P-type semiconductor. The holes in the outer shells due to the three-valence impurity act as majority current carriers*

an electron. The current flows through the material by the holes; as the holes are of a positive nature the crystal is said to be composed of *P-type* material. The P stands for *positive current carriers* because the holes carry the current through the crystal.

Holes are attracted to the negative terminal in the same way that air bubbles move upward in a tube of water—the bubbles move against the force of gravity as they are less dense than the water, and the water replaces them (see Figure A.8).

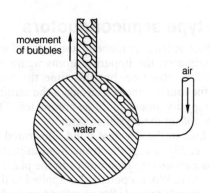

Fig. A.8   *The movement of holes as current carriers in a P-type semiconductor is analogous to air bubbling through water*

The impurity atoms in this example are able to *accept* electrons as the holes migrate to the negative terminal or, if no potential is applied, wander freely throughout the crystal. The three-valence impurity is termed an *acceptor* material.

A.7 Majority and minority carriers

As stated previously, to give the P- and N-type materials the desired conductivity, the amount of impurity needed is very small. To achieve the necessary controlled impurity level, the basic or intrinsic four-valence semiconductor material must be made extremely pure. This level of purity is obtained only after careful and complex refining processes. The ratio of impurities to basic material is less than one part in 100 000 000. Without the desired level of purity of the basic material to start with, and the careful addition of only the exact proportions of impurity, the characteristics of the doped material would make it unacceptable as a semiconductor.

When a crystal absorbs energy in the form of heat or light, the thermal activity releases electrons from the covalent bonds. This then produces both electrons and holes. If the material is N-type, the released electrons add to the normal electron current carriers, and the holes, which are positive current carriers, are in the minority. We say then that in N-type material *electrons* are *majority current carriers* and holes are *minority current carriers*.

The opposite is true of P-type material. Energy-released holes and electrons are added to the normal hole current and so *holes* are the *majority current carriers* and electrons are *minority current carriers*.

It is usual for a minority carrier to last for only a short time, as it can be quickly absorbed by recombination with one of the more numerous majority carriers. This means that in N-type material an impurity hole is soon filled by a free electron and in P-type material an impurity electron soon occupies a hole.

Semiconductor type numbering

Many manufacturers have their own system of type designation codes for the semiconductor devices they manufacture, but the majority use one of two main systems, the American and the European.

B.1 American system

This system is simple but of little use in denoting the semiconductor device type or its purpose. It consists of the prefix 1N followed by a serial number for diodes and the prefix 2N followed by a serial number for other semiconductor devices.

Examples of devices using this system would be 1N1615, which is a silicon diode rated at 5 amperes with a PIV of 400 volts; the 2N3872 which is a 35 ampere 660 volt SCR; and the 2N3732, an audio frequency germanium transistor. In addition to the above the prefix 3N has been given to some FETs.

B.2 European system

Before looking at the present European numbering system, it would be prudent to examine the numbering system in use before 1960. This used the letter O as a prefix to indicate a semiconductor device. A second (and third, if present) letter indicates the general class of device, and a serial number indicates a particular device or development. The second letter(s) use the following code:

A	diode or rectifier,
AP	photodiode,
AZ	zener diode,
C	transistor,
CP	phototransistor,
RP	light dependent resistor.

Examples of this system are:

OA210	semiconductor diode,
OAZ200	zener diode,
OCP70	phototransistor,
ORP30	light dependent resistor.

The present European system is the most logical of any produced. It consists basically of two letters followed by a serial number, and in some cases a suffix denoting a particular type from a family.

The *first letter* distinguishes between a junction and non-junction device and gives an indication of the material type. The following list includes the letters used at the time of writing:

A	devices with one or more junctions using material such as germanium,
B	devices with one or more junctions using material such as silicon,
C	devices with one or more junctions using material such as gallium arsenide,
D	devices with one or more junctions using material such as indium antimonide,
R	devices without junctions using materials such as those employed in light dependent resistors.

The *second letter* indicates the application or construction of the device as follows:

A	low signal diode,
B	variable capacitance diode,
C	transistor for audio applications,
D	power transistor for audio applications,
E	tunnel diode,
F	transistor for radio frequency applications,
G	multiple of dissimilar devices,
H	field probe,
K	hall generator in an open magnetic circuit,
L	power transistor for radio frequency applications,
M	hall generator in closed electromagnetic circuit,
P	light or radiation sensitive device,
Q	radiation generating device,
R	electrically triggered controlling device with high breakdown,
S	transistor for switching applications,
T	electrically or light triggered controlling device with low breakdown,
U	power transistor for switching applications,
X	multiplier diode,
Y	rectifying diode,
Z	zener diode.

The *serial number* consists of three figures for devices designed for use in consumer goods, and one letter and two figures for devices designed for use in industrial or professional applications.

Examples of devices using the above code:

BC108	silicon audio frequency transistor,
AD162	germanium power audio transistor,
BA100	silicon low signal diode,
BY100	silicon mains rectifier (1-2 A 800 PIV),
BYX23	silicon mains rectifier for industrial applications (100 A 1000 V PIV).

As mentioned, a *suffix* is added to the serial number, usually separated by a dash, where there is a variant of a particular family of devices. This mainly concerns zener and voltage reference diodes, rectifier diodes, and thyristors.

For zener diodes and voltage reference diodes the suffix consists of one letter followed by a number which indicates the zener voltage and possibly the letter R. (R indicates that the device is manufactured with reverse polarity, i.e. where the stud is normally the cathode, R indicates that the stud is the anode.) The first letter indicates the nominal tolerance of the zener voltage as a percentage. (A = 1%, B= 2%, C = 5%, D = 10%, E = 15%.) The typical zener voltage is related to the nominal current rating for the whole range. The letter V is used to denote the decimal point when this occurs.

An example of this coding of zener diodes is the BZY88 series, which are silicon zener diodes for industrial applications. A particular type from the range may be BZY88-C9V1. The suffix, C9V1, means that this device has a zener voltage of 9.1 volts with a tolerance of ± 5 per cent.

For rectifying diodes the suffix number generally indicates the PIV of the device and the letter R indicates reverse polarity.

An example of this coding can be made by reference to the industrial rectifier diode discussed earlier, the BYZ23. As the example given had a PIV of 1000 volts, the complete code for this particular diode would be BYZ23-1000. A further example would be the rectifier diode BYX33-1600R. This indicates that it is a silicon rectifier diode for industrial use and it is of reverse polarity (stud is anode) with a PIV of 1600 volts. Incidentally this diode is rated at 400 amperes.

For thyristors, the suffix number indicates either the PIV or the maximum repetitive peak off-state voltage, whichever is the lower.

The BT100A-500R is an example of a silicon controlled rectifier for consumer applications with a peak off-state voltage of 500 volts, with the stud the anode.

In many cases a device may share both an American and European number. An example of this would be the direct replacement of the following pairs of devices in American and European coding:

2N3055 and BDY20, 1N2071 and BY127, 1N712 and BZY88-C8V2.

Resistors

C.1 Preferred values

Almost all components used in electronic circuits are mass produced, with the result that it is impossible to produce each item with exactly the same values. This is common with all components and includes BJTs, capacitors and resistors. The manufacturers of resistors have been able to keep their production to rather close limits and, in general, are able to mass produce resistors to a tolerance of 5 per cent.

A tolerance of 5 per cent would mean that a resistor with a nominal value of 100 ohms could have an actual value anywhere between 95 and 105 ohms. Then if a resistor was nominally rated at 105 ohms it could have the value anywhere between 99.75 and 110.25 ohms. It can be seen from this that it would be a little ridiculous making a 100 and 105 ohm resistor with a 5 per cent tolerance. It is for this reason that a scheme was developed so that a range of resistors in any decade (i.e. between 10 and 100, 100 and 1000, 1000 and 10 000 etc.) could have a 5 per cent tolerance that would not overlap into the tolerance range of the next resistor on the list. This is called the *E24 series* and is based on multiplying each succeeding value in the decade by the twenty-fourth root of ten. In figures, we can say that:

$$^{24}\sqrt{10} \approx 1.1$$

This produces 24 values in any decade. However, in some applications it is not necessary that the tolerance be as close as 5 per cent, and 10 per cent is perfectly satisfactory. In these cases it is possible to use only each second value in the E24 series. This produces the E12 series. Both the E12 and E24 series are called *preferred values* because it is these values which are the smallest number in a decade that will provide the tolerance required and will not overlap in value. Table C.1 lists the E12 and E24 preferred value series.

Table C.1 only lists values in the 10 to 100 decade but the same values are used in any decade. Some examples would be: 0.82, 4.7, 560, 33 000, 1 500 000 etc.

In some special applications a 5 per cent tolerance may be too coarse and in these cases manufacturers can

Table C.1 *Preferred value tables*

10 per cent E12	5 per cent E24
10	10
	11
12	12
	13
15	15
	16
18	18
	20
22	22
	24
27	27
	30
33	33
	36
39	39
	43
47	47
	51
56	56
	62
68	68
	75
82	82
	91
100	100

provide a tolerance of 2 per cent (the E48 series) and 1 per cent (the E96 series). These values are determined in the same manner and consist of multiplying each succeeding value by the 48th root of 10, or the 96th root of ten, respectively. The values in these series need not concern us here.

C.2 Resistor colour codes

Resistors in electronic circuits are coded with coloured rings which designate their resistance value within a certain tolerance. The coloured rings are placed towards one end of the resistor as shown in Figure C.1. The colours

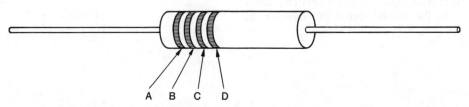

Fig. C.1 *Arrangement of resistor colour code rings. Ring A indicates the first significant figure, ring B the second significant figure, ring C the decimal multiplier and ring D indicates the tolerance limits. (Note that in precision resistors, 1 and 2 per cent, there is one extra ring which denotes a third significant figure, before the decimal multiplier)*

used in resistor coding, together with their numerical values, are set out in Table C.2. Note that the rings are read from left to right, with the rings on the left side of the resistor as in Figure C.1.

Table C.2 *Resistor colour codes*

Colour	Significant figure	Decimal multiplier	Tolerance (per cent)
Black	0	1	—
Brown	1	10	±1
Red	2	10^2	±2
Orange	3	10^3	—
Yellow	4	10^4	—
Green	5	10^5	—
Blue	6	10^6	—
Violet	7	—	—
Grey	8	—	—
White	9	—	—
Gold	—	10^{-1}	±5
Silver	—	10^{-2}	±10

Example C.1

What is the value of a resistor which has rings of yellow, violet, orange and silver from the left?

Yellow is 4, violet is 7, orange means multiply by 10^3 and silver means a tolerance of 10 per cent. The resistor is therefore a 47 000 ohm resistor with a tolerance of 10 per cent.

Example C.2

What is the value of a resistor which has rings of orange, orange, brown and gold from the left?

Orange is 3, orange is 3, brown is multiply by 10 and gold means 5 per cent tolerance. The resistor is therefore a 330 ohm resistor with a tolerance of 5 per cent.

Example C.3

What is the value of a resistor which has rings of blue, grey, green and gold from the left?

Blue is 6, grey is 8, green means multiply by 10^5 and gold means 5 per cent tolerance. The resistor is therefore a 6 800 000 ohm (or 6.8 MΩ) resistor.

Example C.4

What is the colour coding of a 1.5 ohm 5 per cent resistor?

As the first significant figure is 1, the first ring from the left is brown. As the second significant figure is 5,

the second ring is green. As the decimal multiplier is 10^{-1}, the third ring is gold and as the tolerance is 5 per cent, the fourth ring is also gold. The colour code is therefore brown, green, gold and gold from the left.

Example C.5

What is the colour coding of a 56 ohm, 5 per cent resistor?

As the first significant figure is 5, the first ring from the left is green. As the second significant figure is 6, the second ring is blue. As the decimal multiplier is 1, the third ring is black and as the tolerance is 5 per cent the fourth ring is gold. The colour coding is therefore green, blue, black and gold from the left.

C.3 Shortened resistor coding

When used in circuit diagrams, or in list form, it is usual to use an abbreviated form for the value of resistors. Thus, rather than write twenty-seven thousand ohms, or 27 000 ohms, we write 27 kΩ, the k standing for 1000 and the Ω for ohms. Again, instead of saying one million five-hundred thousand ohms, or 1 500 000 ohms, we write 1.5 MΩ, the M standing for million and the Ω for ohms. As normal resistors do not fall into the range of 10^9 or 10^{-3}, only the k and the M multipliers are used.

Another scheme of using an abbreviated form for resistors is a European one that does away with the decimal point and the Greek letter Ω. This was adopted because some printers do not have a font containing Greek letters and in poor printing the decimal point may not show up.

This system uses the three letters R, K and M to stand for ohms, thousands of ohms and millions of ohms, respectively. The letter symbol is placed where a decimal point would normally be. An example would be to represent, in an abbreviated form, the value 6800 ohms: this would be written as 6K8. Similarly, 4.7 ohms can be written as 4R7. If there is no decimal point, the letter multiplier is placed at the end of the numerals. For example, a 27 000 ohm resistor would be written as 27K.

Although this form is sometimes encountered, it is not one which has found a great deal of acceptance here and is not an Australian Standard form.

Boolean algebra

D.1 Why use Boolean algebra?

When working with non-digital equipment and circuits, engineers and technicians use algebra, trigonometry, calculus and other branches of mathematics. In digital circuits all operations depend on either the presence or absence of a signal, and because of this any variable may have only one of two values. This reduces any algebra used to a two-value system instead of a multivalued one.

The mathematics of logic is called Boolean algebra (after its inventor George Boole) and is based on the two-value system. If any description of the operation of a logic system is given in normal language, it must be converted to a Boolean algebra expression before it can be applied to the arrangement of the logic circuits required to carry out the operation. In the Boolean algebra system are postulates, theorems and laws by which expressions may be converted from one form to another. Application of these operations can result in complex systems being converted to much simpler systems and still performing the same logic functions.

These are:

$$0 \cdot 0 = 0$$
$$0 \cdot 1 = 0$$
$$1 \cdot 0 = 0$$
$$1 \cdot 1 = 1$$
$$\left.\right\} \text{AND}$$

$$0 + 0 = 0$$
$$0 + 1 = 1$$
$$1 + 0 = 1$$
$$1 + 1 = 1$$
$$\left.\right\} \text{OR}$$

$$0 = \bar{1}$$
$$1 = \bar{0}$$
$$\left.\right\} \text{NOT}$$

where $\cdot$ means **AND**, $+$ means **OR**, $^-$ means **NOT**.

In algebra a variable is a quantity that can take on the value of any constant in the number system. For example, we can say that 'area equals length times breadth' and we can write this as:

$$A = l \times b$$

In this expression, l and b are variables which may be given any value. Once we give them a value, of any number, the value of A may be determined. In the Boolean system, the only constants (or numerals) which are used are the numerals of the binary system, which are **1** and **0**, so the variables may take only either of these values. From this then we can now list the Boolean algebra theorems.

	AND theorems		**OR** theorems
1.	$A \cdot 0 = 0$	7.	$A + 0 = A$
2.	$0 \cdot A = 0$	8.	$0 + A = A$
3.	$A \cdot 1 = A$	9.	$A + 1 = 1$
4.	$1 \cdot A = A$	10.	$1 + A = 1$
5.	$A \cdot A = A$	11.	$A + A = A$
6.	$A \cdot \bar{A} = 0$	12.	$A + \bar{A} = 1$

NOT theorem

13. $\bar{\bar{A}} = A$

D.2 The theorems and laws of Boolean algebra

Before any of the theorems and laws of the Boolean algebra may be examined, the basic postulates must be confirmed. These come from the definition of the three basic logic gates, the **AND, OR** and **NOT**. When the two states, logic **1** and logic **0**, are considered, the postulates derived from the above gates may be listed.

The laws of Boolean algebra all refer to specific grouping of the variables and imply that where these groupings occur they may be replaced with their equality in each law. Each law is given a name and applies to both the **AND** and **OR** function in most cases. These laws are listed in Figure D.1.

The law of identity	$A = A$	$A = A$
The commutative laws	$A \cdot B = B \cdot A$	$A + B = B + A$
The associative laws	$A \cdot (B \cdot C) = A \cdot B \cdot C$	$A + (B + C) = A + B + C$
The idempotent laws	$A \cdot A = A$	$A + A = A$
The distributive laws	$A \cdot (B + C) = A \cdot B + A \cdot C$	$A + B \cdot C = (A + B) \cdot (A + C)$
The laws of absorption	$A + A \cdot B = A$	$A \cdot (A + B) = A$
The laws of expansion	$A \cdot B + A \cdot \bar{B} = A$	$(A + B) \cdot (A + \bar{B}) = A$

Fig. D.1 *The laws of Boolean algebra*

A	B	$Z = \overline{A + B}$
0	0	1
0	1	0
1	0	0
1	1	0

A	B	$Z = \overline{A} \cdot \overline{B}$
0	0	1
0	1	0
1	0	0
1	1	0

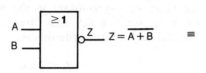

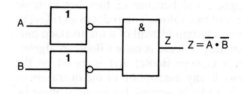

Fig. D.2   *Representation of De Morgan's first law*

D.3   De Morgan's laws

Augustus De Morgan (1806-71), English mathematician, was a contemporary of George Boole. From original research he prepared the way for the rise of modern mathematical logic. His name is commemorated in De Morgan's laws.

De Morgan's first law says that the complement of a sum equals the product of the complements. This can be written in a Boolean equation as:

$$\overline{A + B} = \overline{A} \cdot \overline{B}$$

and can also be represented by two logic gates, each producing the same result in a truth table as set out in Figure D.2

It can be seen from Figure D.2 that the output of each logic gate produces identical results for every case of input. This means that the two gates shown can be interchanged, as they produce the same result.

De Morgan's second law states that the complement of a product equals the sum of the complements. This can be written in a Boolean equation as:

$$\overline{A \cdot B} = \overline{A} + \overline{B}$$

and can be represented by two logic gates each producing the same result in a truth table as set out in Figure D.3. The outputs of the two gates are identical and so the two gates may be interchanged.

With either of De Morgan's laws, rules can be laid down to change an expression to its equality. These rules are often referred to as *demorganisation*. They are:

1. complement the entire function (remove a bar if it is barred, and bar it if it is not barred);

2. change all **ANDs** to **ORs** and all **ORs** to **ANDs** (change all + to •, and all • to +);
3. complement each of the individual variables (remove a bar if it is barred, and bar it if it is not barred).

Example D.1
Simplify the expression $Z = \overline{\overline{A} \cdot \overline{B}}$, using De Morgan's laws.

The logic gate representing the above equation would be:

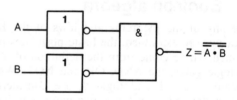

Now, using the steps above the expression will be simplified.

1. $\overline{\overline{A} \cdot \overline{B}} \rightarrow \overline{A} \cdot \overline{B}$
2. $\overline{A} \cdot \overline{B} \rightarrow \overline{A} + \overline{B}$
3. $\overline{A} + \overline{B} \rightarrow A + B$

This should be recognised as an **OR** gate, as set out below:

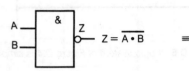

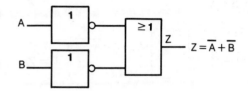

Fig. D.3 *Representation of De Morgan's second law*

D.4 Application of Boolean algebra

The application of the theorems and laws of Boolean algebra are used in the main to simplify logic systems. An example of this simplification can be given.

Example D.2

Simplify the logic system in Figure D.4.

Assume that a logic system has been assembled as in Figure D.4. This consists of three inputs which have been arranged over five inputs to achieve the desired result. The logic expression for these gates may be written as:

$$(A + \bar{B}) \cdot C + (A \cdot \bar{B} + C) = Z$$

The simplification of this expression will now be made using some of the above theorems and laws:

Original expression

$$(A + \bar{B}) \cdot C + (A \cdot \bar{B} + C) \qquad = Z$$

by distributive law

$$(A \cdot C + \bar{B} \cdot C) + (A \cdot \bar{B} + C) \qquad = Z$$

by associative law

$$A \cdot C + \bar{B} \cdot C + A \cdot \bar{B} + C \qquad = Z$$

by commutative law

$$A \cdot C + C + \bar{B} \cdot C + \bar{B} \cdot A \qquad = Z$$

by factor

$$C \cdot (A + 1) + \bar{B} \cdot C + \bar{B} \cdot A \qquad = Z$$

by theorem 9

$$C \cdot (1) + \bar{B} \cdot C + \bar{B} \cdot A \qquad = Z$$

by theorem 3

$$C + \bar{B} \cdot C + \bar{B} \cdot A \qquad = Z$$

(continues over)

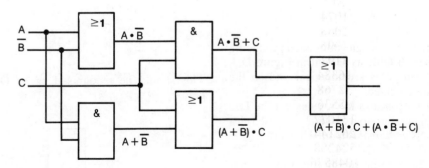

Fig. D.4 *A logic system*

by factor

$$C \cdot (1 + \bar{B}) + \bar{B} \cdot A \qquad = Z$$

by theorem 10

$$C \cdot (1) + \bar{B} \cdot A \qquad = Z$$

by theorem 3

$$C + \bar{B} \cdot A \qquad = Z$$

by commutative law

$$A \cdot \bar{B} + C \qquad = Z$$

It is now quite simple to arrange logic gates that will perform this function and this can be seen in Figure D.5 which performs an identical function as the logic system in Figure D.4

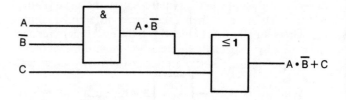

Fig. D.5 *Logic system of Figure D.4 is simplified in this system*

Appendix E

Powers of 2

The following is a list of powers of 2 for use in converting between decimal and binary numbering systems.

2 and power	Expanded number
2^0	1
2^1	2
2^2	4
2^3	8
2^4	16
2^5	32
2^6	64
2^7	128
2^8	256
2^9	512
2^{10}	1024
2^{11}	2048
2^{12}	4096
2^{13}	8192
2^{14}	16384
2^{15}	32768
2^{16}	65536
2^{17}	131072
2^{18}	262144
2^{19}	524288
2^{20}	1048576
2^{21}	2097152

Appendix F

The Greek alphabet

Due to the fact that there are only 26 letters in the English alphabet (which is derived from the Roman one), it can become confusing when a symbol is used for one quantity in an equation, and then used for a different quantity elsewhere. In many instances this is the case, but to reduce the difficulty, especially in electrical quantities, we have made use of the Greek alphabet for many electrical symbols.

While it is not necessary to know in detail where these symbols come from or their names, it could be of benefit to students to have some idea of their origin. Below is a list of the letters of the Greek alphabet, both capital and lower case, their Greek names, and the nearest equivalent sound in English. Not all these letters are used as electrical symbols, but the whole alphabet is presented for students' interest.

Table F.1 *The Greek alphabet*

Name	Capital	Lower case	English equivalent
alpha	A	α	a
beta	B	β	b
gamma	Γ	γ	g
delta	Δ	δ	d
epsilon	E	ϵ	e (short e as in 'met')
zeta	Z	ζ	z
eta	H	η	e (long e as in 'meet')
theta	Θ	θ	th
iota	I	ι	i
kappa	K	κ	k
lambda	Λ	λ	l
mu	M	μ	m
nu	N	ν	n
xi	Ξ	ζ	x
omicron	O	o	o (as in 'olive')
pi	Π	π	p
rho	P	ρ	r
sigma	Σ	σ	s
tau	T	τ	t
upsilon	Y	υ	u
phi	Φ	φ	ph
chi	X	χ	ch (as in 'school')
psi	Ψ	ψ	ps
omega	Ω	ω	o (as in 'hole')

Appendix G

Metric prefixes

In engineering in general, including electrical and electronics engineering, only multiples of powers of three are used to designate multiples and sub-multiples of units. The abbreviations for these power multiples form the prefix for the units. This is seen in microfarad (μF), millimetre (mm), megawatt (MW) and kilowatt-hour (kWh). The following is a list of the metric multipliers encountered in electrical and electronic work, with examples of their application.

The metric prefixes hecto (100), deka (10), deci (10^{-1}) and centi (10^{-2}) are *never* used in engineering. For example, no engineering distance is ever measured in centimetres—this sub-multiple of the metre is only used for clothing and personal measurement.

Table G.1 *Metric prefixes*

Name	Abbreviation	Multiplier	Example
pico	P	10^{-12}	4.7 pF = 0.0000000000047 F
nano	n	10^{-9}	68nF = 0.000000068 F
micro	μ	10^{-6}	10μV = 0.00001 V
milli	m	10^{-3}	120 mA = 0.12A
kilo	k	10^{3}	25 kW = 25 000 W
mega	M	10^{6}	1.5 MΩ = 1 500 000 Ω
giga	G	10^{9}	6.6 GVA= 6 600 000 000 VA
tera	T	10^{12}	8 TWh = 8 000 000 000 000 Wh

Index